Institute on Science for Global Policy (ISGP)

Food Safety, Security, and Defense:
Focus on Food and Water

Conference convened by the ISGP in partnership with
the University of Nebraska-Lincoln, at the Embassy Suites,
Lincoln, Nebraska, United States
October 20–23, 2013

*An ongoing series of dialogues and critical debates
examining the role of science and technology
in advancing effective domestic and international policy decisions*

Institute on Science for Global Policy (ISGP)

Tucson, AZ Office
3320 N. Campbell Ave.
Suite 200
Tucson, AZ 85719

Washington, DC Office
818 Connecticut Ave. NW
Suite 800
Washington, DC 20006

www.scienceforglobalpolicy.org

© Copyright Institute on Science for Global Policy, 2014. All rights reserved.

ISBN: 978-0-9830882-8-8

Table of contents

Executive summary

- **Introduction: Institute on Science for Global Policy (ISGP)**
  Dr. George H. Atkinson, Founder and Executive Director, ISGP, and Professor Emeritus, University of Arizona 1

- **Conference conclusions:**
  Areas of Consensus and Actionable Next Steps 6

Conference program 9

Policy position papers and debate summaries

- *Can We Achieve Global Food Security without Compromising the Use of Water to Meet Other Human and Environmental Needs*
  Prof. Roberto Lenton, Robert B. Daugherty Water for Food Institute, University of Nebraska, **United States** 12

- *Food and Water: A Crisis of Uncertainty*
  Ms. Debbie Reed, Coalition on Agricultural Greenhouse Gases, **United States** 21

- *Improving Livestock Water Productivity*
  Dr. Iain Wright, Animal Science for Sustainable Productivity, International Livestock Research Institute, **Ethiopia** 31

- *Water as an Essential Element in Food Safety*
  Dr. Robert Brackett, Institute for Food Safety and Health, Illinois Institute of Technology, **United States** 40

- *Opportunities and Threats to Widespread Adoption of Bacterial Standards for Agricultural Water*
  Dr. Rob Atwill, Western Institute for Food Safety and Security, University of California, Davis, **United States** 50

- *Improved Water Policies and New Technology will Promote Greater Food and Cellulosic Biomass Production and Reduce Competition for Water*
  Prof. Alvin Smucker, Subsurface Water Retention Technology Program, Michigan State University, **United States** 60

- *Water: A Resource Critical to Food Production and Survival*
  Dr. Elizabeth Bihn, Senior Extension Associate, Department
  of Food Science, Cornell University, **United States** 69

- *Innovation and Policy against Hunger in a Water-Constrained World*
  Dr. Konstantinos Giannakas, Professor and Director,
  Center for Agricultural & Food Industrial Organization,
  University of Nebraska–Lincoln, **United States** 78

Acknowledgment 89

Appendix
- **Biographical information of Scientific Presenters and Keynote Speaker** 92
- **Biographical information of ISGP Board of Directors** 97
- **Biographical information of staff** 101

Introduction
Dr. George H. Atkinson
Founder and Executive Director, Institute on Science for Global Policy
and
Professor Emeritus, Department of Chemistry and Biochemistry and College of Optical Sciences, University of Arizona

Preface

The contents of this book were taken from material presented at an international conference convened by the Institute on Science for Global Policy (ISGP) on October 20–23, 2013, in cooperation with the University of Nebraska–Lincoln, in Lincoln, Nebraska. This ISGP conference, the third in the ISGP program on Food Safety, Security, and Defense (FSSD), focused on Food and Water.

The process underlying all ISGP conferences begins with the recognition that there are significant scientific advances underlying FSSD, a topic that has emerged on the international stage encompassing critical issues affecting the human condition across cultural, ethical, and economic aspects of essentially all societies. Decisions within societies concerning how to appropriately incorporate such transformational science into public and private sector policies rely on candid debates that highlight the credible options developed by scientific communities throughout the world. Since FSSD can potentially have significant impact worldwide, it deserves attention from both domestic and international policy makers from a wide range of disciplines. ISGP conferences offer those rare environments where such critical debates can occur among credible scientists, influential policy makers, and societal stakeholders.

Based on extensive interviews conducted by the ISGP staff with an international group of subject-matter experts, the ISGP invited eight highly distinguished individuals with expertise in FSSD to prepare the three-page policy position papers to be debated at the Lincoln conference. These eight policy position papers, together with the not-for-attribution summaries of the debates of each paper, are presented in this book. The areas of consensus and actionable next steps that were developed by all participants in the caucuses that followed the debates are also presented. The debate summaries and caucus results were written by the ISGP staff and are based on contributions from the conference participants.

Current realities

While the material presented here is comprehensive and stands by itself, its policy significance also can be viewed within the context of how domestic and international science policies have been, and often currently are being, formulated and implemented. While many of our most significant geopolitical policy and security issues are directly connected with the remarkably rapid and profound S&T accomplishments of our time, many societies struggle to effectively use S&T to address their specific challenges. Consequently, it is increasingly important that the S&T and policy communities (public and private) communicate effectively. Recent history suggests that most societies would benefit from improving the effectiveness of how scientifically credible information is used to formulate and implement governmental policies, both domestic and international.

Specifically, credible S&T information needs to be concisely presented to policy communities in an environment that promotes candid questions and debates led by those nonspecialists directly engaged in decisions. Such discussions, sequestered from publicity, can help to clarify the advantages and potential risks of realistic S&T options directly relevant to the societal challenges being faced. Eventually, this same degree of understanding, confidence, and acknowledgment of risk must be communicated to the public to obtain the broad societal support needed to effectively implement any decision.

The ISGP has pioneered the development a new type of international forum designed to provide articulate, distinguished scientists and technologists opportunities to concisely present their views of the credible S&T options available for addressing major geopolitical and security issues.

All ISGP programs rely on the validity of two overarching principles:

1. Scientifically credible understanding must be closely linked to the realistic policy decisions made by governmental, private sector, and societal leaders in addressing both the urgent and long-term challenges facing 21st century societies. Effective decisions rely on strong domestic and global public endorsements that motivate the active political support required to implement progressive policies..

2. Communication among scientific and policy communities requires significant improvement, especially concerning decisions on whether to embrace or reject the often transformational S&T opportunities continually emerging from the global research communities. Effective decisions are facilitated in venues where the advantages and risks of credible S&T

options are candidly presented and critically debated among internationally distinguished subject-matter experts, policy makers, and private sector and community stakeholders.

Historical perspective

The dramatic and rapid expansion of academic and private sector scientific research transformed many societies of the 20th century and is a major factor in the emergence of the more affluent countries that currently dominate the global economic and security landscape. The positive influence of these S&T achievements has been extremely impressive and in many ways the hallmark of the 20th century. However, there have also been numerous negative consequences, some immediately apparent and others appearing only recently. From both perspectives, it would be difficult to argue that S&T has not been the prime factor defining the societies we know today. Indeed, the 20th century can be viewed through the prism of how societies decided to use the available scientific understanding and technological expertise to structure themselves. Such decisions helped shape the respective economic models, cultural priorities, and security commitments in these societies.

It remains to be seen how the prosperity and security of 21st century societies will be shaped by the decisions made by our current leaders, especially with respect to how these decisions reflect sound S&T understanding.

Given the critical importance of properly incorporating scientifically credible information into major societal decisions, it is surprising that the process by which this is achieved by the public and its political leadership has been uneven and, occasionally, haphazard. In the worst cases, decisions have been based on unrecognized misunderstanding, overhyped optimism, and/or limited respect for potentially negative consequences. Retrospectively, while some of these outcomes may be attributed to politically motivated priorities, the inability of S&T experts to accurately communicate the advantages and potential risks of a given option must also be acknowledged as equally important.

The new format pioneered by the ISGP in its programs seeks to facilitate candid communication between scientific and policy communities in ways that complement and support the efforts of others.

It is important to recognize that policy makers routinely seek a degree of certainty in evaluating S&T-based options that is inconsistent with reality, while S&T experts often overvalue the potentially positive aspects of their proposals. Finite uncertainty is always part of advanced scientific thinking and all possible positive outcomes in S&T proposals are rarely realized. Both points need to be reflected in policy decisions. Eventually, the public needs to be given a frank, accurate assessment of the potential advantages and foreseeable disadvantages associated with these decisions. Such disclosures are essential to obtain the broad public support required to effectively implement any major decision.

ISGP conference structure

At each ISGP conference, internationally recognized, subject-matter experts are invited to prepare concise (three pages) policy position papers. For the October 20–23, 2013 ISGP conference in Lincoln, these papers described the authors' views on current realities, scientifically credible opportunities and associated risks, and policy issues concerning Food and Water. The eight authors were chosen to represent a broad cross section of viewpoints and international perspectives. Several weeks before the conference convened, these policy position papers were distributed to representatives from governments, societal organizations, and international organizations engaged with the ISGP (the United States, Italy, the United Kingdom, Canada, Ethiopia, and Brazil). Individuals from several private sector and philanthropic organizations also were invited to participate and, therefore, received the papers. All participants had responsibilities and/or made major contributions to the formulation and implementation of domestic and international policies related to Food and Water.

The conference agenda was comprised of eight 90-minute sessions, each of which was devoted to a debate of a given policy position paper. To encourage frank discussions and critical debates, all ISGP conferences are conducted under the Chatham House Rule (i.e., all the information can be used freely, but there can be no attribution of any remark to any participant outside the conference setting). In each session, the author was given 5 minutes to summarize his or her views while the remaining 85 minutes were opened to all participants, including other authors, for questions, comments, and debate. The focus was on obtaining clarity of understanding among the nonspecialists and identifying areas of consensus and actionable policy decisions supported by scientifically credible information.

The ISGP staff attended the debates of all eight policy position papers. The not-for-attribution summaries of each debate, prepared from their collective notes and recordings of the sessions, are presented here immediately following each policy position paper. These summaries represent the ISGP's best effort to accurately capture the comments and questions made by the participants, including the other authors, as well as those responses made by the author of the paper. The views expressed in these summaries do not necessarily represent the views of a specific author, as evidenced by his or her respective policy position paper. Rather, the summaries are, and should be read as, an overview of the areas of agreement and disagreement that emerged from all those participating in the debates.

Following the eight debates, small groups held caucuses with each caucus representing a cross section of the participants. A separate caucus for the scientific presenters also was held. These caucuses focused on identifying areas of consensus and actionable next steps for consideration within governments and civil societies in general. Subsequently, a plenary caucus was convened for all participants. While the debates focused on specific issues and recommendations raised in each policy position paper, the caucuses focused on overarching views and conclusions that could have policy relevance both domestically and internationally.

A summary of the overall areas of consensus and actionable next steps emerging from these caucuses is presented here immediately following this introduction under the title of **Conference conclusions.**

Concluding remarks

ISGP conferences are designed to provide new and unusual (perhaps unique) environments that facilitate and encourage candid debate of the credible S&T options vital to successfully address many of the most significant challenges facing 21st century societies. ISGP debates test the views of subject-matter experts through critical questions and comments from an international group of decision makers committed to finding effective, real-world solutions. Obviously, ISGP conferences build on the authoritative reports and expertise expressed by many domestic and international organizations already actively devoted to this task. As a not-for-profit organization, the ISGP has no opinions nor does it lobby for any issue except rational thinking. Members of the ISGP staff do not express any independent views on these topics. Rather, ISGP programs focus on fostering environments that can significantly improve the communication of ideas and recommendations, many of which are in reports developed by other organizations and institutes, to the policy communities responsible for serving their constituents.

ISGP conferences begin with concise descriptions of scientifically credible options provided by those experienced in the S&T subject, but rely heavily on the willingness of nonspecialists in government, academe, foundations, and the private sector to critically debate these S&T concepts and proposals. Overall, ISGP conferences seek to provide a new type of venue in which S&T expertise not only informs the nonspecialists, but also in which the debates and caucuses identify realistic policy options for serious consideration by governments and societal leaders. ISGP programs are designed to help ensure that S&T understanding is integrated into those real-world policy decisions needed to foster safer and more prosperous 21st century societies.

Conference conclusions

Area of Consensus 1
Water, a critical resource for food safety and security, requires local, regional, national, and international bodies to prioritize policies and legislation that better manage water use, storage, quantity, and quality.

Actionable Next Steps
- Organize and convene major scientific and policy conferences to identify regional, national, and international water issues and establish agendas for water management at the appropriate level that balance social, ecological, and economic needs.
- Develop a model multi-jurisdictional system (e.g., the High Plains Aquifer Area anchored by Nebraska) that could be emulated in structuring a national water policy.
- Adapt pilot programs and implement incentives (e.g., via the Farm Bill or Water Resources Development Act) to induce all relevant stakeholders to achieve a nationally developed water agenda.
- Pressure government and international agencies to examine opportunities and barriers related to food and agricultural trade associated with water policy.

Area of Consensus 2
Water quality standards, based on scientifically credible risk assessments and cost/benefit analyses, must be tailored to specific uses and locations and reflect environmental and food safety issues, agricultural needs, as well as human and animal health.

Actionable Next Steps
- Conduct scientifically credible risk assessments and cost/benefit analysis, developed by subject matter experts, to establish water quality standards for specific applications (e.g., irrigation, food processing, drinking water).
- Establish policies and incentives to ensure compliance with water quality standards.

- Mobilize relevant communities (e.g., producers, consumers) to provide input to shape the Produce Food Safety Rule with respect to water quality.
- Develop technologies (e.g., genomic applications) to identify, monitor, and control contaminants and pathogens that directly affect human health.

Area of Consensus 3
Application of existing and emerging technologies (e.g., irrigation, drainage, efficient water capture, genetically modified organisms [GMOs], and nanotechnology) is critical for producing a sufficient amount of safe and nutritious food, considering climate change, population growth, dietary changes, and competition for water resources.

Actionable Next Steps
- Develop and deploy technologies that improve water use efficiency and water quality, including integrated agricultural technologies (i.e., organisms that combine traits such as drought resistance with nutritional enhancement), and irrigation and information technologies (e.g., remote sensing).
- Address legal, cultural, and economic barriers (e.g., intellectual property, trade) that restrict both domestic uses and export or trade benefits of emerging technologies is essential to meet growing food demand.
- Incentivize private industry to issue benevolent-use licenses to farmers in less-wealthy countries to maximize adoption of water-efficient technologies.
- Continue to define and harmonize standards for sustainability supply chain initiatives (e.g., water footprint labeling).
- Improve and tailor communication to build trust and enhance public understanding about new technologies and innovations.

Area of Consensus 4
Communication among the scientific community, the general public, farmers, and policy makers is critical to overcoming the lack of understanding of the role of water and farmers in the food system, and the threats to water and food security.

Actionable Next Steps
- Encourage food and water researchers to collaborate with social scientists and communication experts to improve development and delivery of key messages (e.g., social media).
- Create curricula that incorporate water-related issues to ensure children

understand the importance of water issues (e.g., the University of Arizona Project WET) and integrate these programs into core educational curricula (e.g., Common Core Standards in the U.S.).
- Facilitate outreach and communication of Good Agricultural Practices to farmers (e.g., fund cooperative extension, digital communication).

Area of Consensus 5

There is a need to improve the understanding of the relationship between animal production and water use efficiency and quality, including the risks and benefits to human and animal health and the environment.

Actionable Next Steps
- Investigate and understand resource demand regarding non-traditional protein sources for practicality of commercialization and cultural acceptance (e.g., insects, lab-generated protein).
- Develop accurate models of water-use efficiency, quality, and degradation in integrated livestock and cropping systems in the delivery of specific nutrients.

ISGP conference program

Sunday, October 20

15:00 – 17:00	**Registration**
16:00 – 16:30	**Conference Meeting: Science presenters**
16:30 – 17:30	**Caucus Meeting: All presenters and participants**
17:30 – 18:45	*Reception*
18:45 – 19:00	***Welcoming Remarks*** **Dr. George Atkinson,** Institute on Science for Global Policy (ISGP) Founder and Executive Director And **Dr. Harvey Perlman,** University of Nebraska–Lincoln Chancellor
19:00 – 20:30	*Dinner*
20:00	**Evening Remarks** **Prof. Kenneth Cassman**, Professor of Agronomy and Horticulture, University of Nebraska-Lincoln

Monday, October 21

06:00 – 08:45	*Breakfast*

Presentations and Debates: Session 1

09:00 – 10:30	**Prof. Roberto Lenton, Rogert B. Daugherty Water for Food Institute, University of Nebraska, United States** *Can We Achieve Global Food Security without Compromising the Use of Water to Meet Other Human and Environmental Needs?*
10:30 – 11:00	*Break*
11:00 – 12:30	**Ms. Debbie Reed, Coalition on Agricultural Greenhouse Gases, United States** *Food and Water: A Crisis of Uncertainty*
12:30 – 13:45	*Lunch and presentations by* **Dr. Prem Paul**, *University of Nebraska–Lincoln Vice Chancellor for Research and Economic Development, and* **Dr. Ronnie Green**, *University of Nebraska Vice President and Harlan Vice Chancellor for the Institute of Agriculture and Natural Resources at the University of Nebraska–Lincoln*

Presentations and Debates: Session 2

14:00 – 15:30	**Dr. Iain Wright, Animal Science for Sustainable Productivity, International Livestock Research Institute, Ethiopia** *Improving Livestock Water Productivity*
15:30 – 16:00	*Break*
16:00 – 17:30	**Dr. Robert Brackett, Institute for Food Safety and Health, Illinois Institute of Technology, United States** *Water as an Essential Element in Food Safety*
18:00 – 19:00	*Reception at Morrill Hall, UNL campus*
19:00 – 20:00	*Dinner*
20:00 – 20:45	**Evening Remarks** Dr. Andrew Benson, W. W. Marshall Professor of Biotechnology, Director of Core of Applied Genomics and Ecology, University of Nebraska–Lincoln

Tuesday, October 22

06:00 – 08:45	*Breakfast*

Presentations and Debates: Session 3

09:00 – 10:30	**Dr. Rob Atwill, Western Institute for Food Safety and Security, University of California, Davis, United States** *Opportunities and Threats to Widespread Adoption of Bacterial Standards for Agricultural Water*
10:30 – 11:00	*Break*
11:00 – 12:30	**Prof. Alvin Smucker, Subsurface Water Retention Technology Program, Michigan State University, United States** *Improved Water Policies and New Technology will Promote Greater Food and Cellulosic Biomass Production and Reduce Competition for Water*
12:30 – 13:45	*Lunch*

Presentations and Debates: Session 4

14:00 – 15:30   **Dr. Elizabeth Bihn, Department of Food Science, Cornell University, United States**
Water: A Resource Critical to Food Production and Survival

15:30 – 16:00   *Break*

16:00 – 17:30   **Dr. Konstantinos Giannakas, Center for Agricultural & Food Industrial Organization, University of Nebraska–Lincoln, United States**
Innovation and Policy against Hunger in a Water-Constrained World

Caucuses

17:30 – 22:00   **Focused group sessions**

Wednesday, October 23

06:00 – 08:45   *Breakfast*

09:00 – 12:10   **Plenary Caucus Session**
Dr. Matt Wenham, ISGP Associate Director and Dr. Sweta Chakraborty, ISGP Senior Fellow, *moderators*

12:00 – 12:30   **Closing Remarks**
Dr. George Atkinson

12:30 – 13:30   *Lunch*

13:30   *Adjournment*

Can We Achieve Global Food Security without Compromising the Use of Water to Meet Other Human and Environmental Needs?**

Roberto Lenton, Ph.D.
Founding Executive Director, Robert B. Daugherty Water for Food Institute,
University of Nebraska, Lincoln, Nebraska, United States

Summary

One of the most urgent challenges for the 21st century is to achieve and sustain global food security without compromising the use of water to meet other pressing human and environmental needs. Doing so in the face of a changing climate will be extraordinarily difficult and will require scale- and context-specific science and science-based policies and institutions. Research and development will be needed to generate innovative technologies and practices to improve the efficiency and sustainability of agricultural water management and to enhance our understanding of the bottlenecks that constrain the adoption of such innovations. Science-based policies and institutions are needed to (i) create incentives for the adoption of innovative technologies and practices for efficient use of water in agriculture; (ii) control water use to ensure sustainability, especially of groundwater resources; and (iii) manage water storage to help cope with climate extremes.

Current realities

At present, almost a billion people have insufficient food to lead a healthy and active life. While the numbers of people living in poverty have decreased significantly in the last couple of decades and are expected to continue to decrease in the decades ahead, meeting future food needs in light of water constraints will be exceptionally challenging. From the food perspective, overall food requirements will increase at a greater pace than population growth as a result of improving diets resulting from rising incomes. From the water perspective, economic growth and rapid urbanization will result in increased use of scarce water resources, greater levels of water stress, and less availability for use in food production, a situation that will be exacerbated by the impacts of climate change on both water requirements and availability. For these reasons, one of the most urgent challenges for the 21st century is to achieve and sustain global food security without compromising the use of water for other pressing human and environmental needs. Importantly, some of these other needs, especially safe drinking water and sanitation, are vital to ensure appropriate food use and nutrition to the world's most vulnerable populations.

Scientific opportunities and challenges
With regards to water, perhaps the most promising approach to meet this challenge is to improve the management and use of water by and for agricultural and food systems, enabling more food to be produced with less water and less energy. There have been important advances in this area in recent decades, involving innovations in agricultural, irrigation, information, and food processing technologies. Further progress, however, requires better scientific understanding at multiple levels and scale- and context-specific solutions. Research and development will need to focus on generating innovative technologies and practices, such as abiotic stress-tolerant crops, irrigation technology to reduce water losses, and water-smart food processing technologies to reduce water used by the food and beverage industry. Solid social science research is needed to enhance our understanding of the bottlenecks that constrain the adoption of innovations that boost water productivity.

Beyond the use of water, approaches include reducing food waste and increasing agricultural trade to water-short regions. Reducing the significant levels of food waste that currently affect the food supply chain along the full "farm to fork" spectrum, in both more- and less-affluent countries, would help improve water availability and water quality for other purposes and reduce overall pressure on the world's natural resources. The practical and scientific challenges and risks involved, however, are considerable: at the "fork" end of the spectrum, reducing waste involves formidable issues of behavioral change by consumers that will not be easily overcome; while the investments required to reduce waste at the "farm" end of the spectrum are huge. Increasing trade in agricultural products from water-rich to water-short regions and countries (a concept known among water specialists as trade in virtual water) enables water-short countries to save scarce water resources by importing rather than growing food. However, increased trade in virtual water is not a panacea. Although food imports help in ensuring physical access to food, they can negatively affect economic access in countries in which agriculture is the main source of income and livelihoods. In addition, most countries are hesitant to rely too heavily on food imports to address their domestic food needs.

Policy issues
Science-based policies and institutions are required to create incentives for the adoption of innovative technologies and practices for efficient use of water in

agriculture, to control water use to ensure sustainability, especially of groundwater resources, and to manage water storage to help cope with climate extremes.

- *Local authorities and state governments need to create incentives for the adoption of innovative technologies and practices for efficient use of water in agriculture.* Without appropriate incentives, new technologies and practices for efficient use of water in agriculture may never be adopted in practice. In many areas, farmers currently have few strong incentives to conserve water, largely because they do not have to pay for the true value of the water they use. This is a classic problem of externalities, which can be addressed through market-based policies to "internalize" the external costs and benefits to third parties by ensuring that water charges reflect the true costs of water consumed or otherwise impacted by agriculture. In areas where groundwater is the principal source of irrigation water, energy policy can also affect water use; where energy prices are high and the costs of pumping groundwater are considerable, farmers have a strong incentive to reduce water usage and thus pumping costs. The adoption of new technologies and practices for efficient use of water in agriculture also requires appropriate policies to foster the development of supply chains and technical assistance, especially in countries in which such systems are weak.

- *Governments should establish local institutions and policies to control water use to ensure sustainability, especially of groundwater resources.* Good water governance to control and conserve water is needed to ensure sustainability of the resource, in terms of both quantity and quality. Groundwater, as a resource shared by large numbers of current and potential users, is often in danger of being overdrawn, and effective institutions are needed to control its use. Nebraska's Natural Resources Districts (NRDs) provide an excellent example of such institutions. Created in 1972, the NRDs are local watershed-based authorities governed by a locally elected board of directors and with revenue from property taxes and other sources. Over the last 40 years, the NRDs have played a significant role in ensuring that Nebraska's abundant groundwater resources (equivalent to 19 Aswan dams) have remained stable despite the major expansion of groundwater irrigation during that period. Local watershed-based authorities like these, tailored of course to local contexts, might prove helpful in avoiding groundwater depletion in other areas where the groundwater resource is in danger of being overdrawn.

- *Water storage should be appropriately managed to help cope with climate extremes.* Water storage can play a huge role in smoothening out variability in water supply and thus coping with climate extremes (e.g., floods and droughts). Physical storage options range from surface reservoirs and aquifers to wetlands, lakes, ponds, and soil moisture. Decision making around storage can and should involve farmers — the ultimate water managers — as well as local, state, or federal agencies. Farmers can increase soil moisture storage through no-till farming or by adopting drought-tolerant crops that allow roots to dig deeper for water. Public agencies can also use science to increase effective storage, for example by better understanding groundwater recharge and flow processes or predicting reservoir inflows to reduce unnecessary spilling of stored water. Effective water storage management was a major factor in Nebraska's success in coping with the 2012 drought: good groundwater governance by the state's NRDs. coupled with augmented water storage through farmer adoption of drought-tolerant crops and conservation tillage enabled farmers to mitigate the impacts of drought. This kind of integrated approach, tailored as always to local contexts, could help in coping with drought in other situations.

References

International Water Management Institute. (2009). *Flexible Water Storage Options and Adaptation to Climate Change*, Water Policy Brief, Issue 31.

Jägerskog, A. and Jønch Clausen, T. (eds.) 2012. *Feeding a Thirsty World – Challenges and Opportunities for a Water and Food Secure Future. Report Nr. 31.* Stockholm International Water Institute.

Shah, T. (2009). *Taming the Anarchy: Groundwater Governance in South Asia.* RFF Press, Washington DC, 310 pp.

** A policy position paper prepared for presentation at the conference on Food Safety, Security, and Defense (FSSD): Focus on Food and Water, convened by the Institute on Science for Global Policy (ISGP) October 20–23, 2013, at the University of Nebraska–Lincoln.

Debate Summary

The following summary is based on notes recorded by the ISGP staff during the not-for-attribution debate of the policy position paper prepared by Dr. Roberto Lenton (see above). Dr. Lenton initiated the debate with a 5-minute statement of his views and then actively engaged the conference participants, including other authors, throughout the remainder of the 90-minute period. This Debate Summary represents the ISGP's best effort to accurately capture the comments offered and questions posed by all participants, as well as those responses made by Dr. Lenton. Given the not-for-attribution format of the debate, the views comprising this summary do not necessarily represent the views of Dr. Lenton, as evidenced by his policy position paper. Rather, it is, and should be read as, an overview of the areas of agreement and disagreement that emerged from all those participating in the critical debate.

Debate conclusions

- Because water supplies are endangered by opportunistic use, it must be a priority to improve the management of water supplies and optimize its storage to limit, and eventually eliminate major water-shortage issues. Since the challenges associated with water usage and storage generally are context-specific, geographical specific solutions (e.g., at local and regional levels) as opposed to global remedies are more effective.

- While laws regulating water management exist only in some countries, it remains important that legally mandated regulation of water management be implemented broadly. Since enforcement is the principal tool to guarantee proper water management, governments need to ensure such water-use laws are enforced and that violators are prosecuted. When the same water resources involve different countries or regions, negotiations are necessary to ensure that water is properly distributed (not necessarily equally divided) so that the benefits of water are appropriated equitably.

- Increasing the price of water is a powerful tool to reduce usage, distribute supplies to economically sound uses, or reduce waste. The economic paths identified include increased taxation or the internalization of the price of supplying the water resources (e.g., energy) in the final price.

- Since there is the perception that it is too late to reverse climate change and counter anticipated droughts, investments in scientific research are necessary to develop adaptations to water shortages, such as drought-resistant crops, improved water-storage technology, and enhancement of soil moisture.

Current realities

Issues relative to water and food (e.g., waste, shortage, management), are context specific and should be tailored to individual needs. For instance, in some areas in Africa there is a shortage of water, which requires increasing water supplies, while in other areas there is abundant water but relatively high costs create a dearth of irrigation systems, the solution to which requires reducing the costs of irrigation.

In many regions (e.g., North West India, Middle East, Australia) institutions have failed to regulate water usage, especially in those areas of increasing groundwater use and depletion. In some countries, laws have been created but governments have struggled to enforce them, whereas in other regions there is a total absence of rules. Exploitation of water resources by investors in regions that lack regulations and/or enforcement was described as a threat for local populations because they cannot benefit from the natural resources of their own lands.

Good models of multicountry negotiations (e.g., Nile, Danube, and Rhine rivers) were identified as methods to improve regional water usage. However, in most cases, upstream countries/regions/provinces dictate the use of water resources because of the power imbalance existing between upstream and downstream countries/regions. In this context, the lack of multicountry/federal regulations has been identified as the main cause of water usage inequalities throughout regions sharing common water resources.

One of the main causes of water shortage was identified as water-intensive crops (e.g., cane sugar and cotton) being grown in regions of the planet where rainfall is not plentiful, requiring the movement of enormous amounts of water. Decisions on which crops to cultivate are often driven by economic reasons (i.e., more profits from the products) rather than by good water use. One example cited was the water-intensive nut crops grown in Southern California where water resources are scarce.

A concurrent cause of water shortages was identified as drainage. Drainage water (e.g., rain and irrigation water not absorbed by the soil) often is diverted into rivers and oceans instead of being stored as a water supply for subsequent local uses. Recovered floodwaters could be potentially used during drought seasons. Currently, there are few measures to recapture water from drainage and floods.

The link between energy and water usage is being increasingly emphasized and discussed worldwide. Bringing water supplies to more arid regions requires enormous amounts of energy (e.g., fuel to operate water pumps or aqueducts), even though these costs often are not reflected in the price of water.

Scientific opportunities and challenges

It was agreed that global food security could be achieved without compromising water resources for other human needs (e.g., hydration, hygiene, industry) but to do so successfully requires improvement in water management. Often, this is linked to the relationship between the quantity and quality of water (i.e., reducing water consumption frequently leads to a reduction in water pollution).

The debate focused on the importance of considering irrigation and drainage as a whole and not as separate issues. Therefore, the recovery of the drained water, as opposed to its elimination, was considered a potential partial solution to seasonal drought. Developing new techniques to improve the collection of unused water through research emerged as a partial solution to water shortages.

Since appropriate storage of water to prevent flooding would also guarantee water supplies during drought seasons, enhanced water storage could alleviate problems related to both flooding and drought. However, the construction of large dams was not considered a universal solution to the issue but, rather, the need to identify proper, case-specific storage strategies was emphasized.

Water storage optimization was highlighted as an opportunity to ensure the water supply where it is most needed. It is cheaper to store water at the source or directly next to the fields than to build large dams downstream and distribute water from the dam. While there are many useful techniques for keeping water in the fields (e.g., cover crops, tillage practices, and soil moisture management), there is a need to improve the understanding and outcomes of such techniques. In addition, the creation of small wetlands close to the farming areas has been described as a less-expensive remedy compared with the existing strategies for water storage. It was noted that such new wetlands would benefit wildlife as well.

While some considered water management to be context specific (e.g., local), others argued that, in relation to climate change, it is a global issue that requires worldwide collaborative efforts. Attempts to achieve global regulation are challenging because of the varying interests of nations. There was general agreement that attention needs to be focused on developing techniques to adapt to climate changes, because it is likely too late to counter climate changes.

Since decreasing food and water waste would alleviate water shortages worldwide, it is important to recognize that reducing waste largely depends on different factors, such as awareness, culture, and infrastructure. Investing in addressing awareness and changing cultural norms are as important as building more appropriate and modern infrastructure.

Policy issues.
Ground water needs to be managed locally, but negotiations are necessary when aquifers cross boundaries of different regions and countries. There was agreement that mechanisms to balance different needs should be implemented and such mechanisms must focus not on how water is distributed, but on how the benefits of water are shared. The goal of this approach must be the distribution of water resources as is appropriate and not necessarily in equal parts.

Creating incentives for the adoption of technologies and interventions focused on reducing water waste through better management and storage was considered a fruitful starting point for improving water-use efficiency. In the context of improve storage management of aquifers, surface reservoirs, ponds, wetlands, and particularly soil moisture, there was no agreement as to whether water-use control needs to be at the farm, within state or regional institutions, or at the federal level.

International agreements were seen as a potential solution to the management of water usage, especially in those cases involving international river basins or aquifers. A framework for negotiation among parties needs to be developed to create such agreements by broadening the discussion beyond water sharing to include energy, food, and water usage.

It was suggested that improved water management and storage practices would be achieved only by pricing water at the level that it is worth (e.g., by taxation), resulting in reduced and more efficient use. It was agreed that higher costs would induce users to reduce water consumption (probably by reducing water waste). Neglecting to include the costs of transporting water (e.g., energy consumption to operate pumps) encourages the cultivation of water-intensive crops in areas where water is not abundant. Another option for decreasing use and improving efficiency, could be internalizing the costs of energy into the price of water, instead of creating a tax. Both measures in tandem were considered feasible solutions.

The widespread lack of regulations on water use and management in less-wealthy countries was considered a primary cause of speculation by investors, resulting in negative consequences for the local populations. The construction of a legal framework that regulates these speculations was viewed as necessary. It was widely agreed that, from a governmental policy viewpoint, the priority is to ensure those water resources are primarily used to allow access to small holders.

In many countries, enforcement of water use and management laws and regulations, as opposed to the absence of such regulations, is the primary challenge. Regulation at different institutional levels depending on case-specific issues requires that enforcement also be considered and implemented to ensure that existing and future laws are obeyed.

Growing crops in areas where natural resources (e.g., water and soil composition) cannot support them instead of growing those crops where the climate and the soil are most appropriate is driven by the needs of the manufacturing or farming sectors. It was suggested that international trade could be appropriately shaped (i.e., seasonal and local produce favored) by internalizing the costs of natural resources. A system of incentives and regulations could be instituted to ensure that those who are polluting and/or over using water are monetarily affected, thus discouraging the practices

Food and Water: A Crisis of Uncertainty**

Debbie Reed, M.Sc.
Executive Director, Coalition on Agricultural Greenhouse Gases (C-AGG),
Policy Director, International Biochar Initiative (IBI), Washington, D.C.,
United States

Summary

Agriculture in the United States is experiencing structural change, including trends towards larger farms with less diversity of product and less biodiversity, coupled with globalization and the resulting demands of global market forces. At the same time, agriculture is increasingly threatened by uncertainty and instability from global climate change. The impacts of global climate change on water availability alone, in the form of more extreme precipitation events over time (e.g., droughts, floods, and intense precipitation events) will exacerbate impacts of temperature shifts and warming that are anticipated to reduce agricultural productivity over time. The stresses of reduced reliability of water supplies for crops (whether rain-fed or irrigated) and livestock cannot or has not been adequately addressed by policy makers. In fact, farm policies being currently deliberated by the U.S. Congress would subsidize crop insurance for irrigated agriculture more than $500 million more per year than nonirrigated agriculture, at the expense of continued depletion of aquifers upon which many important grain crops in large areas of the country are dependent. While U.S. agriculture has a long history of adaptation to change, it is highly likely the confluence of uncertainty from the aforementioned events will create stresses to the agricultural system that may not allow for adaptation in a timely manner without significant financial, food security, and defense impacts to the U.S. and globally.

Current realities

Agricultural systems globally are increasingly homogenized, specialized, and intensified, resulting in biodiversity loss that threatens global food security. The Food and Agriculture Organization (FAO) estimates that 90% of global food energy and protein comes from just 15 plant and 8 animal species; and that wheat, rice, and corn provide more than half the world's plant-based caloric intake. Water accessibility, quantity, and quality are threatened by increased demands, competing uses, unsustainable overuse and pollution, and the impacts of climate change and agricultural practices. In the U.S., water availability and quality is threatened by oil and fossil fuel extraction practices and the unsustainable depletion of aquifers and ground water. The globalization of markets has created wide swings in food (commodity) prices and availability, driven by food and fuel policies and weather-related events. The complexities and uncertainties of unmitigated climate change impacts and associated weather variability; changes and movement in pestilence, diseases, and their vectors; water accessibility and quality issues; and global competition for food, land, water, and other natural capital creates a global food security crisis of potentially epic proportions.

Climate change models have long predicted impacts that are increasingly observed around the globe. However, climate change models are still subject to uncertainty: they remain fairly coarse, exhibit disagreement between models, and produce at best decadal results of potential impacts that are not region specific. Although U.S. agriculture has long been associated with adaptability to change, the uncertainties of temperature changes and extreme weather events from climate change — including daily and diurnal temperature fluctuations — can stress crops and livestock, reduce productivity, and, driven by vulnerabilities of reduced genetic biodiversity, make regional or larger-scale disasters possible, even in the near term.

The U.S. lacks a federal or unified water policy or law. The Ogallala aquifer, located largely under the U.S. states of Texas, Kansas, and Nebraska, provides a key example of water accessibility issues at stake. Though largely non-renewable, the Ogallala is subject to variable state management policies, laws, and intra-state compacts, and increased demands from urban growth and agriculture. Parts of the aquifer have dropped 80 to 100 feet in the past 15 years. Despite conservation attempts in some areas, droughts have led to increased irrigation, and federal farm policies currently being debated would further subsidize irrigation.

Global market forces impact food supplies and prices; flooding in China in 2013 led to massive failures of wheat crops, which increased U.S. exports by 25 million bushels. Though high production elsewhere helped stabilize global supplies and prices, such incidents in the future could lead to price spikes, food shortages, and global disruptions with economic and security impacts. Water quality impacts from agricultural intensification and nutrient loading have led to localized and regional water quality issues, including algal blooms and dead zones due to hypoxia and eutrophication, with human and wildlife impacts.

A U.S. surge in oil and gas is being fueled by new technologies making previously inaccessible reserves economically accessible. Hydraulic fracturing (aka "fracking") is used in thousands of wells in the U.S. and is expanding rapidly. A single well can use 1 million to 5 million gallons of fracking fluid, the exact contents of which are largely unknown due to proprietary claims by drilling companies. Flow-back water from these wells contains salts, hydrocarbons, heavy metals, isotopic tracers, and other impurities. Because these substances can contaminate groundwater, fracking can aggravate and/or create water shortages.

Scientific opportunities and challenges

Climate-friendly agricultural systems and technologies, and adaption to climate change and water shortages, in the form of drought-resistant crops and crops resistant to rust, other anticipated diseases, and new disease vectors are receiving limited but increased attention. However, the identified uncertainties and variables make it difficult to assess whether the highest priority activities and technologies are being pursued, or whether current policies and activities are appropriate or adequate.

Continued investments in climate-change models to increase accuracy and predictability to finer geographic scales and more real-time or short-term impacts are necessary to help inform policy decisions and prioritize investments and activities relative to climate change mitigation and adaptation, particularly for agriculture. What we do not know is far more critical than what we do know. Long-predicted climate change impacts (e.g., weather and precipitation extremes) are already being observed, but unpredicted anomalies are also being observed, e.g., "stalled" surface temperature increases, which have been variously attributed to volcanoes, ocean sinks, the presence of other pollutants (e.g., sulfur dioxide), or cyclical solar intensity. Just as possible is a reversal or a sudden rapid heating, which might be accompanied by more extreme weather and/or may trigger feedback loops that exacerbate extreme impacts, all of which may overwhelm resources and our ability to adapt — agriculture included.

Science-based modeling of potential pest and disease outbreaks based on known and anticipated climate change and related impacts to crops and/or livestock, as well as the ability to contain and/or treat such outbreaks, is critical to understand food security and related vulnerabilities based on our current over-reliance on genetically homogenous agricultural production systems. Gene banks may preserve critical germplasm, but may not offer solutions in time to avert crises, or to recover from crises in a timely manner.

Policy issues
- Global mandatory climate change policies must be enacted. The United Nations Framework Convention on Climate Change (UNFCCC) process seems broken, and bilateral and multilateral agreements between nations are a band-aid approach. The U.N. should integrate existing overlapping

conventions to address climate change, biodiversity, desertification, poverty eradication, and sustainable development. Economics and legal ramifications add complexity, but should be addressed systematically and by heads of state via the U.N. process. U.S. accession to treaties and conventions is complicated by required Senate passage, and the fact that the U.S. did not ratify the Kyoto Protocol to the UNFCCC or the Convention on Biological Diversity (CBD).

- U.S. federal policies are needed to address water quantity, usage, conservation, and quality. Existing fractured state approaches are inadequate and inequitable legal quagmires. A multistakeholder approach based on a National Academy of Sciences assessment of these issues is warranted, to be led by a neutral third-party facilitator, and include appropriate federal agency representatives, sectoral stakeholders, and the general public.

- International collaboration on water, including a recent High-Level International Conference on Water Cooperation hosted by the Government of Tajikistan (including U.S. State Department participation) and the Shared Waters Program of the U.N. Development Program should continue and be expanded.

- Use and overuse of natural resources such as water and fossil fuels have environmental and public health impacts that do not remain localized. Fracking for oil and gas extraction should be banned and discontinued until its long-term impacts to environmental and human health are thoroughly investigated via a global, scientific, transparent assessment, undertaken by a multisectoral U.N. scientific advisory panel. Full public disclosure of all chemicals used in fracking and cost-benefit analyses of remediation and cleanup are essential components of a credible assessment.

- Sustainable agricultural intensification and food security should be a topic of federal and global efforts, led by the FAO, and must include economic analysis and the internalization of current economic externalities such as the use, impacts, and movement of natural resources and water in the global marketplace. Sustainable supply chain initiatives are circling this issue quite inefficiently, and require global harmonization and standardization of metrics and tools. This assessment should include a realistic approach to match agricultural production to localities and regions best suited to efficient production to maximize efficiencies and minimize inputs. Efficient nutrient utilization technologies such as nitrification inhibitors can help address economic, climate change, and water quality issues simultaneously.

Increased biodiversity (seeds/crops, livestock, soil biota) must be an explicit focus of agricultural sustainability and food security measures; the CBD has laid the groundwork for these efforts.

- The U.S. Congress should remove distorting subsidies from the U.S. Farm Bill. This includes direct payments, $500 million in targeted crop insurance irrigation subsidies, and federal crop insurance subsidies. U.S. Department of Agriculture payments to farmers should be associated with established income caps and require adherence to conservation requirements and conservation plans.

** *A policy position paper prepared for presentation at the conference on Food Safety, Security, and Defense (FSSD): Focus on Food and Water, convened by the Institute on Science for Global Policy (ISGP) October 20–23, 2013, at the University of Nebraska–Lincoln.*

Debate Summary

The following summary is based on notes recorded by the ISGP staff during the not-for-attribution debate of the policy position paper prepared by Ms. Debbie Reed (see above). Ms. Reed initiated the debate with a 5-minute statement of her views and then actively engaged the conference participants, including other authors, throughout the remainder of the 90-minute period. This Debate Summary represents the ISGP's best effort to accurately capture the comments offered and questions posed by all participants, as well as those responses made by Ms. Reed. Given the not-for-attribution format of the debate, the views comprising this summary do not necessarily represent the views of Ms. Reed, as evidenced by her policy position paper. Rather, it is, and should be read as, an overview of the areas of agreement and disagreement that emerged from all those participating in the critical debate.

Debate conclusions

- Since changes in climate and the indiscriminate use of existing supplies have contributed significantly to the depletion of water resources globally, policies must be adapted that accurately consider climate changes (e.g., more frequent and intense floods and drought, temperature increase) as major factors if sustainable uses of water supplies are to be ensured.
- The current models, used to predict short-term climate changes (e.g., intense precipitation, unexpected temperature events), need to be significantly improved since they do not accurately reflect the realistic impact of climate on water supplies. Since the tools employed to measure climate data

(e.g., greenhouse gas emissions and usage of nitrogen-based chemicals in agriculture) are neither routinely standardized nor harmonized, the resultant information, and models on which they are based, do not provide meaningful analyses for predictions.

- The U.S. government's enormous investment (i.e., $500 million per year in the Farm Bill) to subsidize crop insurance and production has resulted in the overproduction of irrigated crops (e.g., corn) and damaged water supplies while benefitting only a select group of producers. A reallocation of this monetary assistance toward supporting nonirrigated crops is urgently required if sustainable water resources are to be available in the future.

- The education of the public concerning the scarcity of water, and its consequences, needs to focus on helping shape public opinion and garner support for policies designed to preserve water resources. The views and priorities of different stakeholders (e.g., consumers, farmers, private industry, and local governments) must be included in decision-making processes that recognize data, obtained from a range of sources that credibly describe the relationships between climate change and the availability of water resources.

Current realities

Although efficient water conservation programs exist, in reality the scarcity of water remains an unacknowledged issue in many localities. It is evident in the U.S. that different states sharing the same water resource do not necessarily undertake similar water-management strategies. For example, the Ogallala aquifer (known as the 1,000 years aquifer for the time required for water replenishment), shared by eight U.S. states, is being significantly depleted by Texas alone (water levels are 15–60 feet below the level from the previous decade).

The absence of standardization in the metrics and tools used to monitor the quantity and quality of water supplies is responsible for the wide range of dsicrepanies among the data measured (e.g., nitrogen use, carbon dioxide emissions). This issue has motivated some multinational corporations to examine the available tools with respect to determining their positive and negative characteristics. The Sustainability Consortium was mentioned as an example of an organization that works to develop transparent methodologies, tools, and strategies designed to facilitate a new generation of commodity items and food products.

The European Union, which provides a positive framework at the federal level, assists individual nations to collaborate in the decision-making processes needed to identify priorities and policies. Member states must adhere to the E.U.-wide regulations and mandates to receive E.U. funding

Governments face serious difficulties in creating unified water policies, mainly because public opinion concerning the severity and immediacy of water shortage are often influenced by the views presented in the media rather than by the collective views of credible scientific studies As an example from the debate on climate change, when the Intergovernmental Panel on Climate Change (IPCC) releases negative news related to climate change, the media routinely gives more credibility to those who disagreed with the IPCC findings. Governments enacting water-usage restrictions in countries without a strong public consensus can create substantial public protest (e.g., Australia).

From the perspective of the private sector, investments in improving the efficiency with which water resources are managed are of critical commercial interest since it directly impacts the availability of water and avoids water shortages that significantly hamper production. Sustainable supply chains for water are essential to ensuring the long-term access to safe, usable water required to bring high-quality products to market.

Hydraulic fracturing, colloquially referred to as "fracking," was considered a threat to water supplies since the water used in this practice cannot necessarily be recycled because drilling companies, citing proprietary reasons, will not disclose the composition of the added chemical mixtures. In fracking, as in the case of genetically modified organisms (GMOs), the processes used and the consequences of their application, are not well understood by the public. However, it was concluded that the lack of transparency in fracking could not be compared to the lack of understanding on GMOs. In the first case, issues concerning the disclosure of information currently dominates while in the second case, there has been an absence of public education.

Scientific opportunities and challenges

As in many topics involving scientific methods, education can serve as a valuable tool for forming sound public opinion on water issues. As in the case of climate change, there is a need to communicate credible scientific understanding to the public concerning the continuing availability of water and its quality relative to human health.

Engaging the public in dialogue to establish the parameters of a sustainable water usage is necessary to ensure public cooperation with the proposed solutions. In this context, soil-moisture sensors and ground sensors have been suggested as good partial solutions to effectively monitor water use.

Measurement metrics and tools need to be standardized and harmonized to ensure that the variables being measured (e.g., carbon emissions) can be consistently compared. Since many tools currently used do not accurately measure the variables of interest, there is a need for the various stakeholders (e.g., multinational corporations and regulatory agencies in governments) to collectively evaluate the existing measurement tools for the purpose of understanding their usefulness. Some tools must be improved and other discarded. For instance, the survey of farmers conducted by the CBD concerning their yearly diesel usage (i.e., their carbon impact) with respect to the carbon footprint of agriculture obtained no reliable data since the tools employed were found not to be reliable.

The Global Alliance on Greenhouse Gases was cited as a good example of a multinational effort to determine research priorities and to determine how resources might be pooled for common uses. Among other topics, the Alliance is studying carbon and nitrogen emissions from land use and the impact of those emissions on climate changes and, subsequently, on water resources and food production. The study demonstrated that climate change is linked to water quality, water quantity, and food security and that an integrated approach is necessary. Past practices were considered to be the reason for the inability to address these issues and to prioritize potential solutions.

The existing models on climate change need to be refined and improved to enable scientists to provide more reliable short-term predictions. While the long-term projections tend to be consistent with each other (i.e., they all foresee global climate changes such as weather and precipitation extremes), the real-time or short-term impacts (e.g., anomalies such as "stalled" temperature increases) need to be more precisely characterized. However, there was disagreement on the utility of models, which some considered "pseudo-science" since these models do not rely on the scientific method of validation of the results.

Farmers do not normally use the current crop models that predict future impact of climate change on crops to inform decisions and manage their crops because these models are not user friendly. The U.S. Department of Agriculture (USDA) released models designed for use by farmers, but the results of each the individual models are not compatible with the other. This incompatibility was considered to be one of the biggest challenges for making any of these models of practical use. Therefore, there is an urgent need to develop crop models that farmers could utilize for better land administration.

The Agricultural Model Intercomparison and Improvement Project developed climate models that "feed" plant-growth models, which in turn "feed" economic models. In this prediction system, while the climate models are the most accurate, the biggest challenge concerning the plant-growth models is the lack of data since private companies are unwilling to release proprietary data.

Determining priorities before an actual crisis occurs is necessary to avoid inappropriate decisions in crisis situations. GMO investment and development was considered essential in countering future food and water shortages. For instance, the development of drought-resistant and salt-resistant crops through scientific research was highlighted as a necessity, together with gene banks, to preserve critical germplasm.

Policy issues

Sustainable water management needs to be addressed through the creation of U.S. federal policies because the existing state approaches (e.g., the Natural Resources Districts) consist of inadequate and unbalanced legal frameworks. Federal policies must be assessed and developed by a respected and unbiased third party (e.g., the National Academy of Sciences) and then evaluated through a multistakeholder process, including representatives of the public, federal agencies, and private sector representatives. This recommendation supports the U.S. Geological Survey's call for a credible scientific body to draw a roadmap with the goal of determining the steps and actions to undertake, given the resources and information available on water issues.

First, current water quality, accessibility, and policies must be understood. Second, the technological innovations, already being used or that potentially could be used, need to be determined and their existing regulatory frameworks identified and evaluated. Third, future actions need to be discussed at the public level.

Given unanimous agreement by all stakeholders (e.g., consumers, farmers, industry, municipalities) on the appropriate course of action, subsequent federal policy on water management must be enforced. It is important that all stakeholders understand and agree upon the regulations that are being implemented. Although federal policies would potentially solve national issues, water scarcity remains a global matter and, therefore, requires international collaborations. To appeal to a wide variety of countries, that otherwise would deal with water problems internally due to sovereignty disputes, it was suggested that the discussions on water be presented in terms of food and energy issues.

The U.S. federal government, through payments released by the USDA to farmers affected by natural catastrophes, is redirecting risk from the insurance companies. Subsidies included in the Farm Bill also have created distortions in the production system (e.g., overproduction of certain crops). Therefore, USDA subsidies must be revised and assigned based on established income caps, which require compliance to water conservation measures.

Improving public education on water quality and usage issues is necessary to prevent the communication mistakes made when GMOs were described to the public. A dual strategy is required: (i) improving science literacy within the public and (ii) communicating the credible scientific understanding of the issues with simple words and concepts that are meaningful to nonexperts. To draw public attention to the topic of water, it is necessary to frame the issue within the context of existing subjects of interest (e.g., health, food, energy). Public attention to a topic will help shape the public support of rational water policies required to implement practical solutions.

Whether hydraulic fracturing should be impeded if not accompanied by full disclosure of the chemicals used during the process remained unclear. One perspective was that fracking should be discontinued until its long-term environmental and public health implications are appropriately described, while another view was that fracking has to be weighed against increasing oil and gas costs and, therefore, is necessary for meet future energy demands.

Improving Livestock Water Productivity**

Iain A. Wright, Ph.D.
Program Leader, Animal Science for Sustainable Productivity,
International Livestock Research Institute, Addis Ababa, Ethiopia

Summary

Demand for livestock products is increasing rapidly in the developing world as a consequence of population growth, rising incomes, urbanization, and changes in dietary preferences. The livestock sector now accounts for about 40% of agricultural gross domestic product (GDP) in developing countries, although the investment in the sector falls far short of its economic importance. While this increasing demand presents opportunities, especially for small-hold farmers, to supply livestock products, it also puts pressure on the natural resource base, including pressure on water resources. Livestock water productivity (LWP) — a measure of the efficiency with which water is used to produce economic benefit from livestock — varies widely among systems, but data on different systems, particularly in developing countries, is scarce. Since water for growing feed is the major water requirement in livestock systems, any measures to improve efficiency of water use for producing feed will have a major effect on LWP. Improving animal productivity also increases LWP, as a lower proportion of the total feed intake (and hence water requirement) is used for maintaining the animal. National governments and international agencies should increase the proportion of agricultural research and development budgets spent on livestock to 20%. This should include providing better estimates of the use of water and LWP for different livestock systems. Major improvement in LWP could be achieved if investments were made in improving livestock productivity through research, development, and extension. Design of irrigation systems needs to include the water demand for production of feed for livestock as well as crops.

Current realities

In the developing world, demand for livestock products is increasing because of population growth, rising incomes (people consume more animal products as incomes rise), urbanization, and changes in dietary preferences. Globally, four of the five highest value agricultural commodities are livestock (milk, beef, and pig and chicken meat) — only rice is higher. More than 1 billion people in developing countries depend on livestock for their livelihoods and livestock account for 40% of agricultural GDP, although investment in the sector from public and private sources at all levels, global, regional and national, is not commensurate with this proportion. For example, in Ethiopia only 10% of recurrent expenditure in agriculture is on livestock. Of the water used for agriculture globally (which accounts for 70% of water used to support human activity) approximately 11% is used for livestock production. Water consumed directly by livestock is less than 2% of this figure, with most water being used for feed production. However, little attention has been paid to policies that could reduce the demand for water to produce livestock products. There are also huge differences in the water requirements of livestock in different livestock production systems — the amount of water used to produce a kilogram of beef from a steer in a feedlot in North America is vastly different from that needed to produce a kilogram of beef from in a mixed-crop livestock small-hold farm in the Ethiopian Highlands. Therefore, applying data from one part of the world to another can lead to erroneous conclusions about water-use efficiency.

Scientific opportunities and challenges

LWP is defined as the ratio of net beneficial livestock-related products and services to the water depleted in producing these products and services. It acknowledges the importance of competing uses of water, but focuses on livestock-water interactions. While LWP is a useful concept, there is a lack of good estimates of LWP for different livestock and mixed-crop livestock systems. It is also known that the range in LWP among systems is huge, mainly due to the 70-fold variation in the amount of water needed to produce forage. Also, most existing estimates of water productivity focus on meat and milk as the main outputs and ignore other uses of animals such as draught power, transport, and the role of the asset value of livestock acting as savings and insurance, resulting in underestimation of true LWP.

Two key strategies for increasing LWP include improving feed sourcing and increasing animal productivity.

Improving feed sourcing

Data suggests that the amount of feed produced from evapotranspired water (i.e. water that is taken up by plants) varies from 0.5 to 8 kg per m^3, which in turn has a major impact on LWP. Livestock systems that utilize crop residues as a major feed source have high levels of LWP as much of the water used in crop growth is used to support grain production for human consumption and the straw or stover is almost a by-product, albeit a very valuable one. One study in Ethiopia shows that as the proportion of crop residues in livestock diets increases from 35% to 70%, LWP increases from about 0.1 to 0.6 USD per m^3. Crop residues are usually of low to moderate nutritive value. Crop improvement programs have focused on increasing grain yield and traits such as disease resistance, but research over the past 10 years has shown considerable potential for increasing the nutritive value of crop residues through plant breeding without compromising grain yield. A 1 percentage point increase in digestibility can increase livestock output by 6% to 8% and increase LWP.

Increasing animal productivity

Animals need feed regardless of whether they produce any product — animal scientists term this the maintenance requirement. Water transpired to produce maintenance feed is a fixed input required for animal keeping whether animals are gaining weight, producing milk, or working. Additional water is needed for production, but since the maintenance requirement is fixed, water productivity increases with increasing animal productivity (i.e., the higher the milk or meat output per animal the higher the LWP). Thus, any measure that increases productivity through better feeding, breeding, or animal health will improve LWP.

Policy issues

- Despite the importance of livestock for food security and poverty reduction, and its significant role in the agricultural economy, the investment in livestock development has been inadequate. National governments and international agencies should increase the proportion of the agricultural development budget for livestock to at least 20%.

- Livestock-water interactions have been mainly ignored in water research and planning. As a consequence, there is limited information on the water requirements of different livestock systems, which vary widely. Data from one system are not applicable to another — just because 15,000 liters of water are need to produce 1 kg of beef from a feedlot in North America does not mean that the same amount of water is needed to produce 1 kg of beef in from a small-holder mixed-crop livestock system in Africa. Researchers need to assess the true water requirements and water productivity in different livestock systems, especially in developing countries. In addition, livestock water productivity estimates must take account of all the outputs from livestock, not just meat and milk.

- There is considerable scope for increasing water productivity by making better use of crop residues for feeding livestock. Crop-improvement programs need to be designed to include traits for increasing the nutritive value of crop residues. Investment in technologies, such as second-generation biofuel technology, could lead to large increases in the amount of animal feed available by breaking down lignin to digestible compounds.

- There is a large "yield gap," the difference between potential and actual level of productivity, in most livestock systems in developing countries. Better animal nutrition, animal health, and breeding would have a large effect on productivity, increase food security, and help reduce poverty as well as improve LWP. National governments and international development agencies need to increase investment in the livestock sector, including in research and development and in effective extension services.

- Few development programs consider the integration of crops and livestock. Development agencies such as the development banks, other international organizations, and national governments need to incorporate water requirements for feed production in the design of irrigation systems. Forage production can consume a considerable proportion of the water but is not factored in to design.

Reference
Peden, D., Tadesse, G. and Misra, A.K. (2007). Water and livestock for human development. Water for Food, Water for Life: A Comprehensive Assessment of Water Management in Agriculture. D. E. Molden (Ed). London, Earthscan and International Water Management Institute, Colombo. pp 485-514.

** A policy position paper prepared for presentation at the conference on Food Safety, Security, and Defense (FSSD): Focus on Food and Water, convened by the Institute on Science for Global Policy (ISGP) October 20–23, 2013, at the University of Nebraska–Lincoln.

Debate Summary

The following summary is based on notes recorded by the ISGP staff during the not-for-attribution debate of the policy position paper prepared by Dr. Iain Wright (see above). Dr. Wright initiated the debate with a 5-minute statement of his views and then actively engaged the conference participants, including other authors, throughout the remainder of the 90-minute period. This Debate Summary represents the ISGP's best effort to accurately capture the comments offered and questions posed by all participants, as well as those responses made by Dr. Wright. Given the not-for-attribution format of the debate, the views comprising this summary do not necessarily represent the views of Dr. Wright, as evidenced by his policy position paper. Rather, it is, and should be read as, an overview of the areas of agreement and disagreement that emerged from all those participating in the critical debate.

Debate conclusions
- Since governments expend less than 10% of total agriculture funds on livestock even though livestock represents 40% total agricultural GDP, government investment in livestock needs to be increased for better, more efficient livestock farming.
- To improve agriculture and livestock productivity, data on actual water use need to be generated for less-affluent countries, where smaller, mixed, and integrated systems of farming are used. Comparing productivity of less-wealthy counties against North American and European large-scale producers is not always achievable or effective.
- Increasing the productivity for animals that provide meat, dairy, and eggs would both meet the growing demand for these products with increased individual animal output and reduced environmental footprint (i.e., fewer animals would require decreased use of valuable resources, such as water).
- The use of genetic modifications and improved crop selection can effectively increase nutritive value and digestibility of crops for animals as well as increase the yield per acre for human consumption; both types of increased productivity would drive more sustainable farming.

Current realities

Livestock is an extremely important commodity around the world as a source of food, labor, and wealth, with an estimated 1 billion people depending on livestock in some way for their livelihoods. Livestock products not only help create a healthier population, but increase the wealth of the people involved. Although livestock accounts for about 40% of agricultural GDP, it attracts only 10% of government investment in agriculture. Raising the investment in livestock by at least 20% can help to close the gap.

Since people are dying of famine at historically lower rates, a major issue currently is nutrition, not merely calories per day. Demand is burgeoning for more nutritious animal products (e.g., meat, milk, eggs) in the developing world because of urbanization and population growth as well as a result of increased wealth, a consequence of which is people want and can afford to consume more animal products. While this increasing demand for animal products represents huge economic opportunities for many more people, it puts a strain on natural resources (e.g., water). The majority of current water use is not for drinking, but for producing food crops and fodder for livestock.

Worldwide, and especially in less-wealthy countries, people tend to produce crops and livestock on small-scale farms with integrated, mixed systems, (i.e., where livestock and crops are raised together instead of on specialized farms). One water policy will not fit all needs; optimal water solutions for North America and Europe will be different than for Asia and Africa. While water use for irrigation (and hence, better crop yields) in Africa has been tied to dam projects on rivers such as the Nile, the primary reasons these dams were built was for hydroelectric power and not for food production. In regions such as Africa, there also are historical and political issues to consider (i.e., the legacy of the British Empire, European colonialism).

Although development of better irrigation systems in less-wealthy countries already is occurring, these advances do not often consider actual water use (e.g., the water required to grow fodder crops instead of food for human consumption). Therefore, the people responsible for developing irrigation projects need to examine relevant cultural and physical issues (e.g., proximity to animals to reduce transportation of crops) of a particular region.

There is a lack of data regarding optimal water requirements for agriculture in less-affluent countries as compared to more-affluent countries (e.g., it is known that it takes 15,000 liters of water to produce a kilogram of beef in North America). Studies have been performed for food crops, but have produced little information regarding water requirements for livestock production.

Scientific opportunities and challenges

Because it is necessary to increase productivity per animal to meet worldwide demand for food, the yield gap can be addressed through efforts combining existing knowledge in better feeding, animal health, genetics, and breeding with new knowledge garnered from technological advances and research. These combined efforts have the potential to result in sustainable improvements in which there is increased productivity, but also consideration regarding efficient water use, greenhouse emissions, and biodiversity.

In more-affluent countries, food waste is generally at the "fork" end, as opposed to less-affluent countries where most of the waste is at the farm end. While there is no current single solution to solve both kinds of food waste, there is the opportunity for genetically modified crops to play an important role in addressing certain kinds of waste (i.e., shelf life) in addition to better yields and improved nutrition.

There is a need to develop cost and impact measurements for water related to livestock, which not only can improve yields per acre, but could potentially ascertain the nutritional impact per water unit. However, there is a significant challenge to create such measurements in areas where small amounts of animal protein can significantly affect human health and wellbeing, because some of those results are difficult to quantify as they involve future earnings and better cognition in school. Water impact per unit needs to take into account other uses of water including hydroelectric power, insurance, and improved growth in children from better nutrition. Water-use impact has been quantified in wealthier countries, but more data are needed from less-affluent countries that use diverse farming systems.

The increased use of crop residues represents a strong opportunity to raise productivity in both crop and animal production. Crop residues generally result from food grown for humans, with the unused parts fed to animals (e.g., stover from rice, wheat, and maize). Animals can graze stover (leaves and stalks remaining in the field following a harvest), or stover can be harvested and concentrated or used for other sustainable measures, such as biofuels. With genetic modification of crops for human consumption, there can be significant improvement in yields, nutrition, and digestibility for animals. This improvement in crop residues could be of special interest to regions like Europe that generally are resistant to the idea of genetically modified foods.

Sequencing of the rice genome has led to a 35% to 47% improvement in digestibility of rice straw with better strains. The most progress has been made in sorghum, which has been bred for increases in both yield and nutrition. There are additional opportunities for improving the yield and nutrition of maize, wheat, and rice, (e.g., more nutritious human food and crop residues, improving yield gaps in Africa and Asia) which could lead to more sustainable global agriculture as the population grows and demand for animal products rise.

The nutritional potential for crop residues may be overlooked because of the lack of communication between livestock and crop scientists. For example, a newly created crop with great economic potential for farmers may be neglected in terms of its potential for increasing nutritional stover crop residues for animals. Better communications regarding research and growth of mixed-use crops for humans and animals, particularly of benefit to less-affluent countries are required to improve productivity.

While there is demand for new technologies (e.g., for the transportation of crop residues) by small producers (e.g., in Africa), there is little information flow between producers and the retail markets as well as a lack of infrastructure to move products to market. Even in countries that make good use of crop residues, challenges remain in handling, storage, and transportation. Smaller feedlots can increase their use of crop residues through improved access to animal fodder. While in less-affluent countries, crops and livestock are much more closely integrated, the challenge remains in providing market access to small-scale farmers. If technology drives down transportation costs and improves access to markets in less-affluent regions, these changes could create farms that focus primarily on one crop, moving away from mixed-use systems and, ultimately, benefitting from enhanced efficiency.

Policy Issues

Building livestock agriculture as a more sustainable part of the food chain will require technological advancements and behavior changes from consumers. There needs to be a convergence of consumption patterns to distribute the economic and health benefits; more-affluent nations need to reduce consumption of livestock products, while less-affluent nations will benefit from more animal proteins.

There is evidence that market forces can influence animal production practices and results (e.g., in Brazil, soaring demand for livestock products prompted an 80% increase in the weight gain of cows over 40 years compared with a worldwide average of 25%). However, such improvements are also subject to the law of diminishing returns as results begin to plateau. In current low-producing systems and regions, there is potential for dramatic improvements in livestock production.

Economics and market factors are important considerations in improving the productivity of small-scale farmers' productivity. Existing models can be adapted to increase productivity for crops and livestock. One such model is Operation Flood, which during the 1970s and '80s made India the largest milk-producing nation in the world through marketing, increasing outlets for milk sales, creating co-ops, and providing better access to veterinary care. For India, the results have almost doubled yield, allowing for enough milk for its growing population while still reducing the number of animals over time, which is a major step toward sustainability.

There is a need to begin measuring actual yield by region and system as opposed to against North American standards (where most of the data currently are generated) to understand realistic yield potential and improve yield gaps (i.e., if a system or region location produces 1.5 kg to 2 kg of milk per day, and it is measured against the actual potential of 20 kg per day in North America, that potential is not realistic for the region). In addition, metrics (e.g., water use) need to be measured in mixed conditions because that is the reality for farmers working in less-affluent countries where small-scale systems of crops and livestock are closely integrated (i.e., farmer raises crops not only to feed the animals producing meat, dairy, and eggs, but to feed the animals used for farm labor as well as farm laborers and family).

It was generally agreed that farm system must be viewed holistically to include livestock, crops, and potential of crop residues. Better scientific outcomes for crop residues requires addressing the gap of communication between livestock scientists and crop scientists as well as the lack of awareness at policy levels regarding the importance of crop residues. While considerations about feeding the human populations must necessarily garner the most resources, some funding has been allocated strictly for fodder crops and, on occasion, has been to the detriment of human nutrition (e.g., as was the case in India).

Since there is pressure from licensed veterinary professional groups to preserve the *status quo* in which only licensed veterinarians are allowed to deliver animal medical care, governments are reluctant to legitimize other types of professionals (e.g., paravets) who can also deliver animal care. There are examples of successful models (e.g., in certain states in India) that involve an overseeing veterinarian with many technicians who are responsible for servicing the animal population. Changes in legislation would be required for such models to be put in place more widely. Other issues included a lack of incentives for veterinarians to service areas that are in need, and a lack of drug production to treat generally unprofitable orphan diseases in animals.

Better access to the marketplace for farmers is needed to improve output in certain countries (e.g., in Ethiopia, farmers are investing in and/or receiving microloans to fund growing additional crops — not to improve output for human consumption, but rather to feed their livestock). The opportunity to access markets for small-scale farmers to both buy and sell products needs to be increased, including reducing trade barriers for import and export. Without access to markets, small-scale farmers cannot afford to adopt the improvements being made in technology or sell their products.

Water as an Essential Element in Food Safety**

Robert E. Brackett, Ph.D.
Vice President, Illinois Institute of Technology; and Director, Institute for Food Safety and Health, Bedford Park, Illinois, United States

Summary

Most public health officials understand that contaminated food and water can be important contributors to both infectious and chronic diseases. What is less understood and appreciated, however, is the degree to which water can impact the safety of foods. Adequately addressing this issue has been, and continues to be, hampered by a variety of factors, including a failure to recognize the importance of water as a food component; a disjointed and poorly coordinated regulatory system; a lack of fundamental scientific understanding of exactly how water can adversely affect food safety; and lack of political will.

Most efforts to address the issue of the water-food nexus focus on the impact of production agriculture and food processing on the availability of water. However, relatively little attention is paid to the fact that, regardless of whether water is used as a direct ingredient in food or used during production and processing, it nevertheless becomes part of the food and as such can be a major risk to the safety of that food.

This realization is only a first step in addressing water's role in food safety. One must then develop and apply appropriate science to minimize the risk, and make necessary water use policy changes to assure that regulatory agencies and the agrifood industry adopt best practices involving water use. However, within the larger issues of water conservation and availability, policy makers need to consider food safety when discussing water use. Water sources often cross state and national borders, and foods are a major component in global trade. Hence, solutions will require international attention and coordination.

Current realities

There is general agreement that water, as a resource, has become a critical global issue. This situation has arisen from an increasing demand at the same time as the world is experiencing a decrease in available water. The most obvious reason for the increased demand is the need for clean, potable water for an expanding global population, expected to reach 9 billion by 2050.

A rising standard of living in many developing countries is also increasing the demand for manufactured products, resulting in a less obvious, but even larger demand for commercial uses of water. The decreases in available and affordable fresh water arise from climactic changes, resulting in drought and salt-water intrusion into aquifers in regions that previously contained fresh water.

Although a variety of users compete for access to water, it is the agriculture and food sector that is the largest user of water, accounting for an estimated 80% of use. Much of this use is attributed to irrigation during production agriculture. However, water is also used during the processing of agricultural commodities into consumer products. Additionally, water is used as a transit medium to move product through the various steps in processing, during the heating and cooling of products, to wash products, and as a direct ingredient in the final consumer product. Consequently, water can be a source of contaminants in foods.

Food safety has become increasingly recognized as an important public health issue. The most recent estimates published by the United States Centers for Disease Control and Prevention (CDC) indicate that as many as 48 million cases, 128,000 hospitalizations, and 3,000 deaths are caused by foodborne illness each year in the U.S. The World Health Organization (WHO) estimates that more than 2 million people die each year from diarrheal diseases and that many of these cases are caused by consumption of contaminated food.

Concern over food safety has received increased attention by policy makers and regulators, and has prompted major regulatory changes in several countries, including China, Canada, and the U.S. In at least one case (the proposed Produce Safety Rule being promulgated in the U.S. under the Food Safety Modernization Act), agricultural water was specifically identified as a primary risk factor to the safety of fresh produce. This and other proposed regulations may adversely impact international trade in foods and perhaps even prompt World Trade Organization challenges.

Despite the recognition of the importance of water in the transmission of human illness, the regulatory system for water is largely disjointed and uncoordinated. In the U.S., bottled water is regulated by the Food and Drug Administration (FDA), whereas municipal water is regulated by local, state, or federal agencies, depending on the location of the source of water. The U.S. Environmental Protection Agency (EPA) sets standards for most uses of water, including potable and recreational uses. In most cases, however, water is regulated as a single component with little regard for what, if any, impact it may have on food safety.

Scientific opportunities and challenges

Acquiring knowledge of how and to what degree water contributes to foodborne illness would be a major move forward in improving the safety of foods. Although the scientific issues surrounding the importance of water to food safety would appear simple, the issue is much more complicated. For example, little data exist on exactly how and to what degree pathogenic microorganisms gain access to foods via water. Without knowing this information, it is difficult to estimate the risk to human health attributed to any particular food, or to develop preventive controls to negate or reduce this risk.

Examples of the fundamental information that is needed include (i) the transfer rate of contaminants from water to food contact surfaces or plant or animal tissue, (ii) the extent to which water-borne microorganisms or chemicals become internalized in plant or animal tissue, and, (iii) credible baseline estimates of concentrations and distribution of potentially hazardous microorganisms or chemicals.

Another important question regarding the quality of water is: how safe is safe enough? An answer to this critical question is needed for policy decisions regarding food and water. Determining the public health impact would require a thorough quantitative risk assessment to ascertain the actual risk presented by typical scenarios.

The primary challenge to initiating research would be logistics due to the diverse sources and location of water, the number and types of contaminants that would have to be analyzed, the compilation of data, and the competing agendas for how the data could and should be used. The latter aspect, having entered the policy and political realm, would need to address leadership and funding as a priority.

Policy issues

Changing the way that the public health and scientific community deal with water as food will require significant policy maker input and would have significant policy implications. Addressing the public health implications of viewing water as food would require at the very least, a high level of coordination between water authorities and those responsible for food safety. In some cases, it would require changes in regulatory authority of one or both of these regulatory entities. Moreover, because food and agricultural commodities are so important to international trade, the issue would need to be addressed globally. Success in improving the safety of foods by addressing water issues would have both public health and societal benefits. Not only could it reduce the potential for foodborne illness, but it would also reduce regulatory burden by allowing regulatory agencies to focus more on prevention efforts, rather than detection and investigation of outbreaks. Improved prevention of foodborne illness would result in increased confidence in governments and trading partners.

- Concern over water has largely focused on conservation and availability of clean water. Although this concern is clearly warranted, it does not fully convey the role that water plays in food safety. It is critical that public health officials, regulators, and food companies understand, acknowledge, and appreciate that in order to produce safe foods, the quality of water used in the process is critical. There needs to be agreement that water must be treated as a national and international food safety issue.

- The disparate manner with which water has been viewed and considered globally has led to a situation in which fundamental issues involving water, and resulting interactions between the various uses of water, has never been fully considered. As one initial step forward, the National Academies of Sciences and the Institute of Medicine should be charged with conducting a comprehensive study of the public health impact of water as food, with the goal of identifying critical missing and researchable data.

- The public health community has largely underappreciated the role of water in food safety. Consequently, the contribution of water to food safety issues has likewise not been a priority for research funding. To that end, major funding agencies including the National Institutes of Health, U.S. Department of Agriculture (USDA)/National Institute of Food and Agriculture, and the National Science Foundation should be encouraged to include the topic as a major priority and fund resulting projects accordingly.

- Governments should consider placing regulatory authority for water intended to be used in agriculture for food within one agency. Although individual federal agencies responsible for food safety, primarily FDA and USDA/Food Safety and Inspection Service, would be good candidates to house such authority, a cabinet level department that would coordinate the efforts of all regulatory players might also be considered.

- Because water use and quality crosses national borders, the issue of water's role in food safety should also be addressed by international science and public health bodies. In particular, the WHO and Food and Agriculture Organization should convene a study group to determine how water resources can be used and shared without compromising agricultural products and foods in which it is used. The result could be a White Paper that serves as guidance for national and state efforts.

** *A policy position paper prepared for presentation at the conference on Food Safety, Security, and Defense (FSSD): Focus on Food and Water, convened by the Institute on Science for Global Policy (ISGP) October 20–23, 2013, at the University of Nebraska–Lincoln.*

Debate Summary

The following summary is based on notes recorded by the ISGP staff during the not-for-attribution debate of the policy position paper prepared by Dr. Robert Brackett (see above). Dr. Brackett initiated the debate with a 5-minute statement of his views and then actively engaged the conference participants, including other authors, throughout the remainder of the 90-minute period. This Debate Summary represents the ISGP's best effort to accurately capture the comments offered and questions posed by all participants, as well as those responses made by Dr. Brackett. Given the not-for-attribution format of the debate, the views comprising this summary do not necessarily represent the views of Dr. Brackett, as evidenced by his policy position paper. Rather, it is, and should be read as, an overview of the areas of agreement and disagreement that emerged from all those participating in the critical debate.

Debate conclusions

- Because water's role in the transmission of foodborne illness is a direct threat to human health, one agency, or an interagency task force, needs to be charged with ensuring the safety of both food and water. At a regulatory level, water is currently treated as a discrete element, separate from food, with many different agencies involved that do not communicate effectively enough to ensure the safe use of water in both human and animal foods.

- To contain the costs related to implementing technologies that facilitate clean water for agriculture, global standards need to adopt a cost-risk analysis that considers the many variables that impact water quality, especially for water used in the process of growing and handling food. Since changes to global standards currently are not fast enough to be effective, representatives from Codex member countries need to more rapidly disseminate the relevant information. Of special importance are data describing how water contaminates food before consumption, which are factors that directly affect human health and illness in less-affluent countries.

- While negative incentives (e.g., consumer lawsuits) for food producers and processors generally ensure higher-quality products, positive government incentives (e.g., product labels noting higher safety standards) also need to be developed to encourage farmers and processors to use high-quality water at all stages of food production.

- Since consumers demand food safety information that is often compatible with the lay person's understanding, it is difficult for the public to accurately assess food contamination. Therefore, it is important for authorities to consider how information released may be result in adverse public reactions (e.g., public panic). Educating the public about food safety and ensuring transparency in how food is produced and from what sources is important to avoid adverse public reaction while providing an appropriate level of sophisticated detail concerning food safety.

Current realities

The role of water in the transmission of foodborne illness represents a direct threat to human health, but remains an underappreciated factor to policymakers. Although the WHO estimates that up to 2 million people die each year from diarrheal illness stemming from food and perhaps more specifically, from the contaminated water in food, it remains difficult to identify the specific contributions of contaminated water. In the United States, it is estimated that up to 3,000 deaths a year are caused by contaminated water, and organizations such as the U.S. CDC provide estimates developed from statistical models, but not direct data sources. However, questions remain about the veracity of global statistics on death attributed to inadequate quality and quantity of water because of the difficulty in separating the real causes, times, and sources of water contamination in the food production process.

At a regulatory level, water is currently treated as a discrete element, separate from food. Water is an essential element that is relevant to several agencies governing many spheres (e.g., food, disease, agriculture). In the U.S., water standards are set by the EPA, food is regulated by the FDA and USDA, and diseases are monitored by the CDC. Because water is an integral component of food production, its management requires social, political, and economic commitments among public policy groups, regulatory agencies, and the scientific community.

Water is essential at several points of agricultural production (e.g., washing produce and equipment, irrigation, field cooling and heating, processing, use in flumes to physically move products). A small amount of water contamination has the potential to contaminate the rest of the water supply and subsequently the entire process for which the water is being utilized (e.g., growing, harvesting, and bringing food to market).

A risk-based approach to regulation is increasingly being implemented in more-affluent countries where uniform food-safety standards are established. However, because certain safety standards can be expensive (e.g., treatment of water to make it potable), lower standards throughout water production may be acceptable in less-affluent countries. Consequently, policies regarding food safety can vary considerably across countries.

More-affluent countries have introduced water safety standards (e.g., in the U.S., the Food Safety Modernization Act of 2011 has a clause that water safety standards be added two years after its passing, which has yet to be enacted). More-affluent countries also have the technology (e.g., large-scale producers can recycle and clean their water) to ensure high-quality water throughout all stages of production. However, companies still need to properly use the technology (e.g., a listeria outbreak from a company that recycled its water, but failed to properly chlorinate after that process).

Cost is an important factor to consumers when buying food. Nutritious food is becoming more expensive and in certain countries (e.g., U.S.), it is often cheaper to buy fast food than fresh produce. It was argued that imposing further technological standards would not reverse this trend. However, nutritious food does not always have to be fresh produce, and canned vegetables and fruits are sometimes even more nutritious and have the advantage of having a longer shelf life.

Currently, most incentives for the production of high-quality products are negative and come from the market economy. Grocery stores will reject products from certain factories and farmers if they feel uncertain about its quality. Large companies (e.g., Wal-Mart) have leverage because of their buying power, and food producers are thereby incentivized if they want to keep selling their products to a given customer. Additionally, successful lawsuits from consumers regarding contaminated food incentivize producers to maintain high safety standards.

Scientific opportunities and challenges

It was agreed that in a perfect world, all water sources used to grow and produce food meant for human consumption would be of high standards at the very beginning of the process, from irrigation to the grocery store. While the technology is currently available to make this happen, the challenge is that it is prohibitively expensive to implement. Due to this expense, there is a need for decisions and standards to be set at international and regional levels, depending on a cost-risk ratio on which each region or nation is willing to compromise.

Quality, not just quantity, of water has to be considered in discussions of food safety, and the opportunity exists to set scientific quality standards for water use in agriculture. The value of investing in clean water at the beginning of the agriculture process could far outweigh the costs of disease outbreak. Using uniform high drinking water standards as early as the irrigation process is a goal, but not yet achievable in less-wealthy countries. Risk-based standards are required based on data that measure the exposure and types of contaminants (e.g., heavy metals and microorganisms) that can have chronic and acute effect on public health. In less-affluent countries, any improvement can often have significant impact on human health, whereas in more-affluent countries a change of standards may only have a negligible impact on public health.

Regional variables in the agriculture process include risk tolerability, which is influenced by sociological considerations such as wealth, culture, history, as well as scientific data such as infection rates, likelihood of exposure to pathogens during the entire farm to fork system, and essential differences in regards to the types of produce harvested. For example, different kinds of produce may be cooked at a later stage or eaten raw, while others are cooked during processing or frozen. More information is needed to address the regional differences (e.g., considerations for proximity of contaminants to open-air irrigation sources would be different for food eaten fresh or with little processing than something to be highly processed at a later stage).

The technology exists to clean almost all underground, surface, and recycled water to an acceptable level if countries are willing to endure the costs. Whether the current volume of safe, pre-treated water exists to enact clean water irrigation standards remains under debate. The only way to keep costs down is to create different water standards for the different food production stages (e.g., different water standards are required for the irrigation stage as opposed to the processing stage), and considerations need to be taken on how and when the food will be consumed. Also, many countries still irrigate with sewage or waste water, but more data are needed on the use of sewage water in irrigation to increase public trust. Although many people in more-affluent nations have a negative reaction to the idea of the use of treated sewage water in food production, there is no current evidence that there is a significant negative impact on human health.

A real opportunity lies in improved, more rapid methods to identify contamination, especially with fresh foods, before they leave the processing facility. Identification can be difficult because processing and shipping foods with high moisture content must be rapid, as these products spoil faster, leaving little time to test the products for safety before shipping. Because of this danger, it was agreed that more focus should be on preventive technologies as opposed to reactive ones.

Although it was agreed that most produce becomes contaminated on the surface of the plants during washing, handling, or transportation, questions, especially from consumers, remain as to whether a plant can be contaminated during the growth stage from sewage or recycled water. Controlled studies have shown that this is possible, but is not a statistically significant concern to public health. The general consensus is that handling and processing remain the major sources of food contamination, but there is a need for further study and improved communication of those findings to consumers.

Water safety and security was discussed in relation to military and government-sponsored production and byproducts. For example, perchlorates from rocket fuel production and arsenical pesticides can contaminate water sources. Concerns were also raised about other human-produced chemicals, heavy metals in water, and hydrofracking. While wealthier countries generally have a good understanding of safe levels and treatments of these contaminants, there is a need for more baseline studies and the development of less expensive methods to retain a high level of safety in treated and reused water.

Policy issues

It is essential to address food and water safety in tandem because the quality of water used in food production and treatment affects the safety of the food consumed. The current complex regulatory environment charges many agencies with responsibility for regulating many different elements (e.g., food, medicinal products), all of which rely on water. Therefore, it was recommended that one agency be charged with ensuring food and water safety, or that an interagency task force is created to address these issues. Some called for a subcabinet level coordinator or "czar." There were calls for one overarching agency to deal with water in the U.S., although it was argued that this proposal for a new separate agency or massive reorganization of already existing institutional arrangements is not feasible. If the concept for one overarching agency is not implemented, existing agencies need to consistently communicate with each other on issues related to water use in both human and animal foods.

Codex, developed by the WHO, is an international set of standards on basic food safety (e.g. microorganism and chemical contaminants), which is considered a good attempt at global harmonization. However, Codex member countries are required to introduce any new changes in standards, a process that moves slowly. Representatives of member countries must be made aware of food and water issues in a timelier manner and have the opportunity to raise any relevant concerns quickly.

Questions were also raised about the ability of the private sector to self regulate at an acceptable level. There was agreement that governments need to set regulatory standards, but the private sector needs to contribute its own safety data. Whether data contributed by the private sector can be trusted remains under debate because of uncertainty behind private sector motivations (e.g., political, ethical).

Food producers already self-impose, voluntary standards (e.g., having workers follow sanitary procedures, protecting local surface water sources, implementing radio-frequency identification tag systems to ensure traceability of each batch of food). These standards can be further incentivized through positive means (e.g., government permission to label products with an official seal from companies or farmers proven to be safe in the same way some foods are labeled as organic).

Authorities are under tremendous pressure from the public to communicate information about food safety. Timely reports on possible contaminations are appreciated by consumers, but there is poor public understanding of the science and risks behind food and safety standards. Because the public is unequipped to analyze the information, members may not behave rationally (e.g., unfounded public panic). While it was agreed that consumers deserve transparency and education about their food sources, the onus lies with the food producers to have high safety standards because, even with education, consumers cannot be expected to have the expert knowledge needed to understand the statistics behind food safety reports.

Opportunities and Threats to Widespread Adoption of Bacterial Standards for Agricultural Water[**]

Edward R. Atwill, D.V.M., Ph.D.
Director, Western Institute for Food Safety and Security,
School of Veterinary Medicine, University of California, Davis, California,
United States

Summary

Within the proposed Food Safety Modernization Act (FSMA) Produce Safety Rule are new standards for the maximum allowable concentration of the indicator bacteria, *Escherichia coli*, in agricultural water used on covered crops. This will be first time that microbiological standards will be imposed on millions of acre feet of water for many growers across the United States. Numerous opportunities and challenges will occur as the Food and Drug Administration (FDA) and grower community endeavor to implement these regulations, requiring new science and new policy to facilitate the process. A key concern will be the lack of alternative water supplies for many regions of the U.S. and a lack of proven, affordable treatment technology for the large volumes of raw agricultural water. Without access to loans or cost-share mechanisms, low resource and smaller farming operations may find it difficult to comply. Moreover, growers may find it difficult to implement conservation programs under these new standards, such as reuse of irrigation water, which can concentrate microbiological contaminants. Nine policy action items are proposed to facilitate implementation of these new FDA standards, including federal and industry funding to jump start development of new treatment technologies, requiring irrigation districts to also comply with these new standards, conducting scientific research into more efficient sampling methods, and reaching regulatory consensus between FDA's efforts to improve water quality and the Environmental Protection Agency (EPA) and U.S. Fish and Wildlife Service (USFWS) conservation plans.

Current realities

During the past decade there has been a growing list of produce commodities associated with multi-state outbreaks of foodborne illness, ranging from leafy greens and cantaloupe to berries and almonds. As a consequence, the microbiological safety of produce has become a key focus of FDA's FSMA, expressed in part as the proposed Produce Safety Rule. Within this rule are proposed new standards for agricultural water that are based on U.S. EPA microbiological standards for recreation water, most notably the reliance on standards for the concentration of the indicator bacteria, *Escherichia coli*, that signal when agricultural water is of sufficient quality for its intended use.

The proposed rule defines agricultural water as "water used in covered activities on covered produce where it is intended to, or is likely to, contact covered produce or food-contact surfaces, including: water used in growing (including irrigation water directly applied, water used for preparing crop sprays, and water used for growing sprouts) and in harvesting, packing, and holding (including water used for washing or cooling harvested produce and water used to prevent dehydration)." This comprehensive definition indicates that across the United States the microbiological quality of millions of acre feet of agricultural water will be closely regulated. For example, according to the Farm/Ranch Irrigation Survey (U.S. Agricultural Census), in 2008 there were 54.9 million acres of irrigated land in the U.S., with 91.2 million acre feet of water applied to crops per year. In 2005, about 58% of the total volume of irrigation water used in the U.S. was from surface water sources according to the U.S. Geological Survey. Surface water derived from agricultural watersheds is notoriously vulnerable to microbiological contamination due to a variety of biotic, abiotic, and anthropogenic processes. Hence, adoption of these standards for agricultural water will undoubtedly result in considerable financial and operational difficulties for thousands of growers across the U.S., particularly for low-resource farmers and in regions of poor water quality.

Scientific opportunities and challenges

There are numerous scientific and policy opportunities and operational challenges to the implementation of the proposed microbiological standards for agricultural water. A key challenge is the reliance on using the standard of 126 *E. coli* bacteria per 100 ml of water. This threshold for the maximum allowable concentration of bacteria for compliance will be exceeded for many sources of surface irrigation water across the U.S., particularly when water is impounded or reused under water conservation plans. Sources of water at risk for noncompliance would be irrigation ponds in the Southeast, sediment and tail-water basins in the arid West, and low gradient streams in mixed use (animal and plant) agricultural valleys of the Midwest. Climate such as summer thunderstorms that generate surface runoff typically result in spikes in bacterial indicators in local rivers and lakes, resulting in acute failures of compliance. If a farmer fails to comply with the bacterial standards, then he or she will be required to cease using the source of irrigation water and either treat and test for compliance, or switch water sources. One can imagine scenarios such as the source of irrigation water being a large river, such as the Yakima River in eastern Washington used to irrigate apples, or the All American Canal in the Southwest U.S. Federal and state natural resource agencies such as EPA and the U.S. Department of Agriculture (USDA) Natural Resources Conservation Service (NRCS) may be reticent to permit chemical disinfection of surface water from such sources as rivers, streams, lakes, and ponds. For example, for growers that depend on the Colorado River for irrigation, treating such a large body of water is unfeasible. In these cases, the farmer will need to install treatment technology at the off-take or point-of-use, or reduce the upstream source(s) of bacterial contamination. Reducing these upstream sources can be technically challenging, requiring months to achieve compliance, not to mention issues of private property rights and the need for excellent watershed-scale landowner cooperation. If treatment or prevention options are limited either due to cost of technology, excessively large volumes (e.g., Mississippi River), lack of landowner cooperation, or local restrictions on chemical intervention (e.g., fisheries), then switching sources of irrigation water may be feasible in some circumstances. If the farmer has only a single well, canal, or other source of agricultural water, then acute failures of compliance with bacterial standards can lead to acute loss of access to water, a potential problem in the arid areas, low-resource, or small-farm regions of the U.S.

The other key scientific opportunity is the dearth of data demonstrating how the proposed bacterial standards for agricultural water promote the microbiological safety of produce for all agricultural systems and growing conditions across the U.S. These standards will likely indicate when substantial deterioration of agricultural water quality has occurred, but widespread scientific evidence exists that these bacterial standards can result in a false sense of security. There are many examples of drinking or recreational water meeting bacterial standards, yet outbreaks of waterborne enteric disease occurring or enteric pathogens being present. The scientific and regulatory communities are encouraged to fast track research and innovation in the area of effective waterborne indicators for produce microbial safety and to generate credible datasets that prove which water monitoring strategies and pathogen detection technologies are cost-effective and closely aligned with promoting produce safety.

Policy issues
- Farmers will need technical assistance to comply with the bacterial standards for agricultural water. To facilitate adoption of the FSMA Produce Safety Rule, it is recommended that FDA, USDA, and the World Health Organization (WHO) conduct a thorough and unbiased international

review regarding proven technologies that can affordably reduce foodborne pathogens and bacterial indicators in high volumes of raw agricultural water, to be completed by June 30, 2014. The technical report will be distributed to commodity organizations, agricultural agencies, state and country farm bureaus, and cooperative extension services, and provided to countries that export produce commodities covered by the Produce Safety Rule to the U.S.

- Based on critical technology gaps identified from the technical review, the scientific community (academic, government, private) must immediately expand work on developing chemical, thermal, physical, and other processes that can affordably reduce foodborne pathogens and bacterial indicators in high volumes of raw agricultural water. Funding would be provided by state and federal agencies for basic science, and produce commodity market orders for proof-of-concept projects as needed.

- Depending on location, irrigation districts control much of the irrigation water available to growers. FDA is to implement policy by June 30, 2014 that requires irrigation districts to comply with the microbiological standards for the product they sell to growers (i.e., irrigation water) when alternative water supplies are not available.

- Effective immediately, USDA should shift resources to the NRCS and Cooperative Extension Service to implement on-farm practices and conduct grower workshops for implementing water quality compliance programs and how to remediate water quality problems, with similar programs to be implemented by allied grower associations.

- Uncertainty exists regarding the efficacy of indicator *E. coli* to accurately signal the microbiological safety of agricultural water. To improve the accuracy of water monitoring, testing should be performed for pathogens when a regional water source has a history of detections for a specific pathogen. For example, *Salmonella enterica* is a consistent adulterant of tomatoes from southeastern U.S., hence farmers should monitor for *Salmonella* in conjunction with indicator *E. coli* in these agricultural systems. In the event a pathogen is detected in agricultural water for covered produce, FDA and state agencies need to develop policy regarding the disposition of the exposed crop (e.g., test and hold, or destroy the product).

- The costs of implementing microbiological standards for agricultural water are due in part to the frequency of sampling. Alternative strategies are needed to improve monitoring efficiency, such as the scientific community and FDA assessing alternative sampling strategies that focus more on critical periods (e.g., last two irrigations prior to harvest) and increase the volume of water per assay (1000 ml) to better represent the microbiological quality of water. Highly secure water supplies (e.g., deep wells) and installment of FDA-approved treatment technologies would qualify the grower for reduced sampling frequency. This should be completed by June 30, 2015.

- In regions of shared water resources (e.g., western U.S. federally funded irrigation projects), the agricultural community can implement cooperative sampling for networks of growers served in common by these systems. Cooperative data will therefore represent the source of water for all participating growers.

- There will be inevitable conflicts between grower efforts to comply with the microbiological standards and water or wildlife conservation plans advocated by agencies such as the NRCS. NRCS, FDA, EPA, USFWS, and vested state agencies must identify key areas of potential conflict and work to develop regulatory agreement on how growers can achieve water quality compliance when competing conservation regulations are in place, ideally by June 30, 2015.

- Environmental pressure to reduce agricultural use of surface water in regions with critical fisheries habitat (e.g., Klamath River in western U.S.) will pressure growers to move to more efficient irrigation systems (drip) and water reuse infrastructure (tail-water ponds). Tail-water ponds are unlikely to comply with the proposed FDA microbiological standards, but USDA, EPA, and USFWS could subsidize the cost or provide low-interest loans to growers in critical aquatic habitat to shift from water-intensive to water-conserving irrigation systems.

** *A policy position paper prepared for presentation at the conference on Food Safety, Security, and Defense (FSSD): Focus on Food and Water, convened by the Institute on Science for Global Policy (ISGP) October 20–23, 2013, at the University of Nebraska–Lincoln.*

Debate Summary

The following summary is based on notes recorded by the ISGP staff during the not-for-attribution debate of the policy position paper prepared by Dr. Rob Atwill (see above). Dr. Atwill initiated the debate with a 5-minute statement of his views and then actively engaged the conference participants, including other authors, throughout the remainder of the 90-minute period. This Debate Summary represents the ISGP's best effort to accurately capture the comments offered and questions posed by all participants, as well as those responses made by Dr. Atwill. Given the not-for-attribution format of the debate, the views comprising this summary do not necessarily represent the views of Dr. Atwill, as evidenced by his policy position paper. Rather, it is, and should be read as, an overview of the areas of agreement and disagreement that emerged from all those participating in the critical debate.

Debate conclusions

- Current testing methods and technologies for detecting pathogens (i.e., samples taken regularly at given intervals) must be improved to focus on high-risk events (e.g., prior to harvest, post heavy rain, flooding) and to be effective in examining multiple samples. The costs of these new tests and technologies must be balanced with potential social and cultural impediments.
- From a cost-benefit perspective, determining an acceptable level of risk from pathogens in drinking water is preferred to achieving zero risk (an option available with existing technology). The total absence of pathogens is not necessary to ensure human and animal health, especially since it is evident that repeated small dose exposures creates immunity and a complete absence of water-borne pathogens may create an environment conducive to a catastrophic failure to prevent a pandemic.
- Protecting small farmers from the negative impact of more restrictive water-safety standards can be best achieved by providing the funding (e.g., credits, low-interest loans) needed to implement the safety standards and effective testing rather than exempting small-farmers from regulation. Such regulation must also consider the major differences in farms and farmers (e.g., size of production, types of products, different regions).
- The highly distributed process termed farm-to-fork involves numerous stakeholders with different priorities, but with a shared responsibility for the

integrity of the overall process, including food and water safety. Regulatory decisions must reflect these different relationships (e.g., livestock farmer pollution affecting produce farmers) to be effective.

Current realities

International standards for food safety must recognize the disparity between more- and less-wealthy nations. Effective solutions must recognize the context within which they are applied as well as specific priorities (e.g., the priority for one nation may be basic safe drinking water, while in wealthy nations such as the U.S., safe drinking water is not a well recognized problem). It was noted that the WHO developed a set of standards related to pathogen contents in water, but the standards were considered too lenient by certain countries (e.g., U.S.), and the WHO ultimately dropped them. Further, no consensus has been reached regarding existing acceptable standards that could be authorized globally. The U.S. is in the process of imposing new standards through the FSMA, but the effectiveness of these standards remains to be determined.

It was argued that attempting to impose international standards has caused food safety issues to be in competition with food security issues, a situation with the potential for unintended consequences. In an attempt to comply with standards, farmers in less-wealthy countries may overuse limited drinking water needed for human consumption.

Although zero tolerance for pathogens in water is an achievable goal technologically, it is almost prohibitively expensive. It is more realistic from the perspective of the cost-benefit analysis to determine acceptable levels of risk for water purity. Such a cost-benefit balance can be measured terms of monetary costs and/or with respect to social values (e.g., rural versus urban needs, fish and game hunters versus wildlife preservation advocates). There are, of course, a wide range of "ecosystem services" to balance the desire for untainted food and potable water against the willingness to accept certain levels of risk. Critical, candid dialogues between various groups and communities are needed to address their differing values and acceptable levels of risk for water quality.

Legal considerations have hindered the adoption of new technologies (e.g., genomic sequencing) given public concerns over the potential for microbiological hazards. For example, genomic sequencing can help to identify the specific strain of salmonella that causes sickness in humans. However, because of the potential negative impacts on business, large agricultural companies may not want a specific pathogen to be traced back to specific companies or growers. It was suggested that the relationship between large agribusinesses and commercial laboratories is too close and may result in a conflict of interest in terms of identifying contaminations.

Scientific opportunities and challenges

It is difficult to ascertain the percentage of disease outbreaks that are indisputably linked to contaminated water. Therefore, a reassessment is required to ascertain if conducting a variety of tests over a period of time would be more cost effective, and particularly if different risk assessments may be more effective for different regions. It was argued that it is not yet known with certainty when or how in the agricultural process a specific food becomes contaminated (e.g., ground water versus surface water) and therefore, new tests need to be developed to identify and distinguish different types of water contamination.

Innovative technologies for testing (e.g., utility of biochar) will be driven commercially while and expanded regulations will result from the promulgation of more restrictive water standards. "Strategic sampling," based on monitoring during events having elevated risk is one important technological approach (e.g., testing for pathogens and their decay before food harvesting). There are also methods to test more efficiently that are more useful in mitigating food safety risks (e.g., potential infection and temperature abuse during transportation and consumer end use) than simply testing at given intervals (e.g., monthly). However, because of costs, these new testing methods and technologies may be implemented more readily by larger growers that small-scale producers.

Questions were raised about the efficacy of individual standards for pathogens (e.g., E. coli) with respect to the real danger to public health (e.g., what is the relative effectiveness of tracking an individual pathogen versus tracking all microbial events?). These questions related to the general issues of the cost effectiveness associated with monitoring larger or multiple samples from different locations after specific events as opposed to collecting a sample at given intervals (e.g., monthly). Since it is known that risks from pathogens increase after heavy rain or flooding, standard procedures need to mandate testing after such events. Certain pathogens and food products pair with frequency (e.g., farmed foods are more prone to salmonella and E. coli while risks for listeria are higher during processing). It was agreed that limits on pathogens should be set at a conservative level, and treatments with ultraviolet radiation and increased filtration are needed to meet drinking water standards.

Although in some countries regulatory agencies can help enforce standards, it was argued that smaller farmers are too diverse and unorganized to be regulated and that incentives are a more effective method of ensuring food safety. Large distributers (e.g., Wal-Mart) with the financial leverage to buy and sell the products are able to demand that certain criteria and standards be met. Such incentives create opportunities to alter the traditional tax payer-funded model leveraging one element of the food industry against another to encourage self regulation. Countries will likely trust certain multinational brands or importers (e.g., Walmart) to consistently provide safe products.

Advances in technology will require that consumers be better educated to understand and respect the information on labels. It was noted that it is unlikely that consumers would follow all the rules and instructions, resulting in increased risk, but perhaps lower costs.

Policy issues

Placing the ultimate responsibility for water safety on the farmer would only serve as a negative incentive for the farmer and potentially result in businesses or other stakeholders taking fewer safety precautions. Small farmers may also be negatively affected by the passing of water safety standards (e.g., FMSA), because they are not considered exempt due to the size of their total farm. Accountability and legislation regarding water safety must be reexamined to avoid unintended negative consequences (e.g., undue stress on small farmers).

The U.S. FDA has estimated that the distribution of $460 million a year among farmers to adopt water standards would lower cases of illness related to contaminated food by 1.75 million cases and result in $1.04 billion in savings. Therefore, rather than exempting small farmers, the solution is perhaps to impose the safety standards equally among all farmers, and then provide credits, low-interest loans and other financial assistance to the small farmer. Other suggestions included allowing small neighboring farmers to team and share the cost of implementing the safety standards (e.g., testing the same water source, encouraging groups of small farmers to cooperate to improve food quality and testing).

Intensive livestock agriculture and fracking contributes to pollution, and it is important to ensure that small produce farmers are not solely held responsible for water safety. It was noted that livestock and produce farming face unique economic and ecological challenges (e.g., when the livestock farmer pollutes, it affects the produce farmer down stream). It was suggested that the regulatory burden be distributed across water infrastructure, while protecting the unique features of the different irrigation districts utilized by many farmers (e.g., the Central Arizona Project [CAP] is an open air canal that where pathogens can drift in).

Positive incentives for food safety were suggested that included allowing food products to bear a certificate or seal that indicates that it was grown, cleaned, and packaged with clean water sources. While many farmers groups have tried different kinds of certifications (e.g., organic labels), a credible, independent, third-party audit is required for legitimacy. While these certifications can often be stricter in more-wealthy countries than globally accepted regulatory standards require, most food manufacturers will not pay a premium to the farmers for "better" products. Rather, willingness to pay a premium stems from consumers who desire these products, and an importing nation may trust the certification for trade purposes. There was no consensus regarding how to address increasing cost of operations due to consumer demand without higher values on crops. In addition to market forces incentivizing premium labeling, there are other elements that are not related to food production that can be used as incentives for farmers (e.g., good ecological practices like recycling water, fair labor conditions, and wildlife refuges), which are recognized by certain consumers as desirable. Wildlife refuges in particular are a point of contention where buffers of natural land are preserved and kept from agricultural use. However, there is concern that these conservation practices may actually be the source of some pathogens, and attempts to save land may be moving faster than the science to test them.

Preventative technology must be the focus for ensuring safe water, but some countries do not accept certain types of technology (i.e., ultraviolet radiation in the United Kingdom). Cultural and social differences must be addressed through education and communication efforts to ensure that the benefits of preventative technologies may be fully reaped.

Improved Water Policies and New Technology will Promote Greater Food and Cellulosic Biomass Production and Reduce Competition for Water**

Alvin Smucker, Ph.D.
Professor of Soil Biophysics; Director, Michigan State University Subsurface Water Retention Technology Program, Michigan State University, East Lansing, Michigan, United States

Summary

Growing world populations, increased transport costs, and associated greenhouse gas (GHG) contamination, combined with changing climates demand the establishment of national and multinational water policies, designed for adoption by local regions and landscapes. Newly established region-specific policies need to be coupled with water pricing, best plant/soil management practices, and temporary subsidy support to those who adopt new water-saving science and technologies. Policies are needed for enforcing graded reductions in water use among domestic and industrial sectors in urban and surrounding rural regions during prolonged reductions in annual precipitation. Graduated water-pricing increases, based on essential and aesthetic water volume use, need to be coupled with reported water conservation practices by domestic, industrial (including agricultural), and city government users located across urban centers and rural regions. Historically, many urban centers and farmlands were located along or near rivers, primarily for access to water sources and transportation. Therefore, new policies are required to address how rivers are used in metropolitan, industrial, and agricultural areas to minimize the contamination of these resources. Unlimited water use by agricultural communities also requires expanded environmental water use policies. In the United States, competing federal agencies need to develop a national water policy that establishes guidelines focused on maximum water-use efficiency and minimum water pollution, which include improved guidelines of the Clean Water Act. Historic regional water policies need to be replaced by community based policies developed and implemented by local authorities. This approach should be designed to incorporate systems approaches that consider local/regional, social, political, economic, environmental, and agroecological practices that ultimately lead to the policy adoption. Individual, corporate, municipal, and state organizations adopting these new community-based policies should receive government funding as they incorporate new technological tools that lead to annual water conservation. The world anxiously awaits technological solutions that overcome or at least diminish these more frequent droughty conditions associated with changing climates.

Current realities
Historic approaches for resolving agricultural water deficits have included the global expenditure of hundreds of billions of dollars, invested in hydroelectric and irrigation water reservoirs with the goal of distributing surface water through networks of canals. Dozens of billions of dollars have been spent developing bioengineered plants to improve drought tolerance. There has been a rapid expansion in the agricultural and horticultural irrigation industries, using competitive and often unsustainable quantities of surface and groundwater sources. Often, large quantities of irrigation water are applied to soils unable to retain adequate quantities of water for the plant to achieve maximum production of food and fiber. We are rapidly approaching limits of food production, where many of these historic policies and approaches need to be improved by using new technologies to retain more water where it falls.

Agricultural irrigation practices are often in competition with urban and industrial water needs. Surface water accumulations in reservoirs are expensive, are breeding grounds for disease-carrying insects and animals, and place surface water vulnerable to maximum evaporation. Additionally, 10% to 40% of this surface water evaporates during storage and transport in open channels before it reaches the droughty soil and is absorbed by plant root systems. Nearly all fertile soils requiring little to no supplemental irrigation for sustainable agricultural production are already farmed. Therefore, options for expanding food production in the U.S. and most countries require better management of soil water in the root zones of plants, especially those growing on marginal lands. Increasing food production by 60% to 70% to meet the food needs for a human population approaching 9 billion by 2050, with current limited water resources, requires more efficient soil water storage. Increasing irrigation of sands and other marginal lands, mandates new technologies to increase soil water-holding capacities. Such efforts will transform these areas into sustainable agricultural production in closer proximity to urban centers.

As new sustainable water conservation technologies emerge, current water use for maximum food production paradigms require policy changes that better manage fresh water losses to the oceans from many of our large fresh water lakes. Some of these retained water volumes could be used for supplemental irrigation of food crops. This will sustainably produce more food and fiber, sequester more carbon, protect the environment, and provide more nutritious sustainable food value chains to large urban centers. This paper offers new and revised policies that could resolve many of the current restraints for maximizing food value chains with less water.

Scientific opportunities and challenges

Current water policies in the U.S. and most countries do not permit, nor do they enable, satisfactory solutions to the growing human competition for our most valuable natural resource: **nonsaline fresh water.** Since agriculture uses approximately 70% of this valued natural resource, soil scientists, hydrologists, plant scientists, and engineers need to discover and develop new water-conservation technologies that increase water-use efficiency. New subsurface water retention technology (SWRT) has been developed to convert billions of sandy soil acres into sustainable agricultural production systems that double grain and biomass production with less water.

Greater population, combined with improved diets containing more protein, will require up to 70% more food production. Prescription irrigation that includes greater soil water and nutrient retention in plant root zones, when combined with new policies, offers new approaches for producing more crop per drop of water. We have proposed a trilogy of integrated low-cost technologies that address weaknesses in current farming practices, which require less irrigation when coupled with the best integration of these technologies. These new technologies can be installed with little maintenance for continuous operation in a manner that transforms agriculture and elevates domestic income. Overcoming short- and long-term water deficits for agricultural crops is a key step toward maximizing newly developed hybrids, associated pest management, and protection of harvested produce. Although estimates of food insecurity vary it has been suggested that feeding this many people requires incremental changes in both water technology and water policy.

Policy Issues

- National water policies need to include just and uniformly useful water laws that supersede the plethora of state and regional laws and practices.
- The U.S. Environmental Protection Agency (EPA), Department of Interior, Department of Energy (DOE), Department of Agriculture-Natural Resources Conservation Service, as well as other federal and state agencies need to formulate national water policies that establish guidelines focused on maximum water-use efficiency and minimum water pollution. These new policies must include rewards through tax incentives and temporary subsidies for those who lead, as well as taxing irrigation food production

systems, industries, corporations, and individual dwellings that are inefficient and unsustainable.

- These new federal and state water policies need to be coupled with community-based policy development, implementation, and litigating boards. Only highly efficient irrigation systems of lawns, gardens, parks, sports areas, and agricultural crops should be permitted to operate. Community-established policies that adjust to wet and dry seasons should tax inefficient water use by industries, municipalities, and all irrigation sources.
- Governments must identify the most strategic locations for best irrigated dryland agricultural production and develop permits for the use of public surface waters and renewable deep water reservoirs.
- State and regional governments need to establish seasonal water consumption parameters for households, industries, and agricultural needs in specific regions that pollute rivers, lakes, and streams.
- Government and crop insurance companies must support farm crop losses only when the most water use-efficient irrigation systems are coupled with drought-tolerant cultivars planted on farm lands located in arid and semiarid regions.
- Limit major river flow to oceans by establishing more accurate controls of water levels in the five Great Lakes and other large water storage bodies resulting in available water for on-farm irrigation of grain crops.
- General population support of the best water use for food production should include certification of food production that includes most water-use efficiency rankings and food transport distances on food labels.
- Federal and entrepreneurial initiatives are needed for balancing shipping traffic, electrical energy production, and fresh water irrigation using water from rivers in the U.S. and globally.
- Reduce conflicts and current water wars by establishing free to low-cost water-use policies with surrounding countries that parallel and complement current free-trade policies.

References

Ash, C, Jasny, B.R., Malakoff, D.A., Sugden. (2010). How to get better yields, soil fertility, more crop per drop, better seeds, pest free. *Science,* 327: 808-809.

Bush, S.R. et al. (2013) Certify sustainable aquaculture? *Science*, 341, 1067-1068.

Federoff, N.V. et al., (2010). Radically rethinking agriculture for the 21st century. *Science*, 327: 833-834.

Smucker, A.J.M. and Basso. B. (2013). Global Potential for a New Subsurface Water Retention Technology — Converting Marginal Soil into Sustainable Plant Production. In: G.J. Churchman (Ed.) *The Soil Underfoot: Infinite possibilities for a finite resource.* Chapter 24. CRC Press.

> ** *A policy position paper prepared for presentation at the conference on Food Safety, Security, and Defense (FSSD): Focus on Food and Water, convened by the Institute on Science for Global Policy (ISGP) October 20–23, 2013, at the University of Nebraska–Lincoln.*

Debate Summary

The following summary is based on notes recorded by the ISGP staff during the not-for-attribution debate of the policy position paper prepared by Dr. Alvin Smucker (see above). Dr. Smucker initiated the debate with a 5-minute statement of his views and then actively engaged the conference participants, including other authors, throughout the remainder of the 90-minute period. This Debate Summary represents the ISGP's best effort to accurately capture the comments offered and questions posed by all participants, as well as those responses made by Dr. Smucker. Given the not-for-attribution format of the debate, the views comprising this summary do not necessarily represent the views of Dr. Smucker, as evidenced by his policy position paper. Rather, it is, and should be read as, an overview of the areas of agreement and disagreement that emerged from all those participating in the critical debate.

Debate conclusions

- Significantly improved federal water policies are needed that provide a framework to assist local/regional/state governments to design more uniform regulatory guidelines that are readily adaptable to the specific needs of individual localities regions, and states. For instance, national guideline and policies concerning maximum water usage efficiency and minimum pollution can facilitate community-based decisions on how to regulate case-specific interventions (e.g., lawn irrigation).
- Since water shortage is not perceived as a global issue, especially by those in countries that have not yet faced any relevant consequences, informed

public opinion needs to be fostered in advance of crises. Such heightened public awareness can be achieved by involving the media, schools, and communities, and by requiring the certification of food production that includes water-efficiency rankings and food transport distances on food labels.

- New technologies designed to increase water-use efficiency, SWRT can retain water at the root areas of plants, These technologies need to be implemented at the global level to maximize water resources and combat its shortage by converting sandy soils into highly productive media by optimizing water consumption (e.g., the same production level per acre can be obtained while using half of the water normally used).

- Measurements to determine the efficiency of recycling and water recovery from usage and atmospheric agents need to be implemented through the establishment of local/regional regulations. For instance, drainage water from irrigation can be used again for the same purpose, thus limiting water losses and pollution of the receiving systems (e.g., rivers, lakes, oceans). Moreover, dams and reservoirs can be built to receive water from floods and heavy rainfalls, which can be utilized during drought seasons.

Current realities

The public perception of water shortages greatly varies by geography; for example, the Great Lakes is a region in the U.S. where water is abundant and as a result, people often abuse water as opposed to U.S. states such as Arizona and Nevada, where water is viewed as a high-priced commodity. Therefore, even though the shortage of water supplies is a global issue, regions/countries that have plenty of water tend not to recognize the need for adopting water-saving procedures in contrast to regions where water is viewed as a commodity (e.g., the city of Las Vegas pays its residents $1.50/square foot to replace grass-covered yards with stones or drought-tolerant plants to reduce water use).

There was debate as to whether water from rain and floods should be stored for drought as opposed to replenishing the rivers and, thus, their deltas, which act as a buffer for salinity. The Indus and Seine rivers are examples of ecosystems that are being endangered by the increasing salinity of the water resulting from the penetration of seawater. However, it was noted that water routinely drained from the adjacent agricultural fields into the rivers contains huge amounts of contaminants (e.g., nitrates and pesticides) that promote the formation of anaerobic environments and toxin production.

In many cases, water is overused even for aesthetic purposes, such as for the excessive irrigation of golf courses (e.g., in Florida) to compensate for evaporation and to maintain the proper salinity of the soil. Concurrently, crops are being irrigated with high-quality water at levels higher than needed by plants because the composition of the soil cannot retain sufficient water for plant growth.

Wetlands have been created close to some rivers (e.g., Mississippi River) to accept floodwaters and store them for future uses. While it is difficult to manage the amount of water that enters these systems, strategies have been adopted to control the amount of water release. Similarly, in other regions (e.g., Nebraska), tile lines have been created to drain water during high-water seasons so that water can be easily absorbed by the soil. New technologies are being used to control river water levels and irrigate back through the tile lines depending on the season.

Water currently is not being recognized as a global concern as compared with other issues (e.g., air quality). However, in many regions (e.g., Southern France, Spain, Northern Africa), the public is aware that their respective countries are facing water-shortage problems. Yet irrespective of this, the governments are not adopting appropriate measures to address water shortages. Spain was cited as one example of this controversial behavior because it is still charging the lowest irrigation rates in Europe.

Scientific opportunities and challenges

The increasing population of our planet (9 billion by 2050) will be accompanied by the necessity of increasing global food production by at least 70%, which will result in increased water usage and a subsequent increase in farmland drainage into rivers and lakes. Therefore, the challenge for agricultural practices and the development of scientific options is to minimize the losses of drainage water by adopting proper farming techniques (e.g., no-till practices, cover crop, terracing) and by discovering and developing new methods for water retention.

Existing water policies do not facilitate solutions to the increasing demand for water. Since agriculture uses 70% of this natural resource for irrigation, there is an opportunity for scientists to develop novel water-conservation technologies to increase water-usage efficiency through interdisciplinary efforts (e.g., plant and soil scientists working closely with engineers). A new method, SWRT, has already been developed, and could be used to convert sandy soils (currently 16 billion to 18 billion acres) into sustainable agricultural fields Such methods could result in greatly increasing agricultural and biomass production while using less water. Part of the as-produced biomass can be used to generate organic matter for the soil, while part of it can be used for other purposes (e.g., bio-ethanol production). SWRT has the potential to combat droughts because it is capable of retaining water in the root area of plants for at least two weeks.

Prescription irrigation has been proposed as a valuable tool for reducing water usage and waste and, concurrently, for maximizing the efficiency of irrigation. By integrating technologies that require less irrigation, water is retained in the root area of the plants, enhancing production while saving water resources. These new technologies can be installed at costs comparable with those that farmers are currently using to tile drain wetlands for agriculture, while being extremely durable and efficient. This ensures long-term monetary savings from infrastructure building and, concomitantly, an increase in earnings through enhanced crop yields.

There was agreement on the need for water recycling. For instance, water used for irrigating lawns or crops should be recovered and used again for the same purpose, thus reducing the amount of chemicals used yearly because the recycled water already contains those chemicals. This would greatly reduce water usage and waste in the U.S., where, for example, 38% of the water purified to drinking water standards is being used for flushing toilets. This would eventually involve different piping systems as well as consideration on how to charge for the quality of water delivered. The city of Oakum, Michigan, was cited as an example of a municipality that promotes water recycling and reduces water consumption by providing customers with two different water sources: one specifically for irrigating the lawns and one for other uses.

Urban agriculture, the growing of plants and the raising of animals within and around cities, in areas where water supplies are abundant has been suggested as a partial solution to the water-shortage issue. Increasing this practice would reduce the amount of water being delivered and lost because of evaporation during transportation to farmlands as well as lower the costs derived from shipping.

Policy issues

It is only by anticipating a crisis that the negative effects can be effectively mitigated. Accurate information regarding water-shortage issues is needed to help shape the degree of public awareness and action required to avoid the consequences of droughts. To achieve this, the public must be informed through the media, schools, and community. Awareness of water shortage could also be achieved by introducing water-use efficiency labels on consumer products, similarly to what is required for nutrient compositions.

In the U.S., competing federal agencies (e.g., the EPA and DOE) need to formulate national water policies, including general water laws (e.g., seasonal water consumption parameters) that set a framework for local/regional/state regulatory practices, most of which are specific needs and limitations. Models of successful regional regulation (e.g., in three counties in the southwestern Michigan, a limit has been put on the drilling of new wells because the pull rate is faster than the recharge rate) can be utilized as examples where only highly efficient irrigation systems of lawns, parks, sports areas, and farmlands are allowed.

Community-established policies that adjust to wet and dry seasons must be enacted to encourage efficient and responsible water usage through tax saving benefits and to discourage water waste and pollution through tax increases (e.g., double rates), irrespective of the nature of water consumption (e.g., industry versus farm).

There is also a need to modify existing regulations on wetlands (established by the Natural Resources Conservation Service within the U.S. Department of Agriculture) which prohibit the use of accumulated water. A usage limitation of wetland water should exist for protection of the wetlands, but the drainage of excess amounts deriving from rainfalls must be allowed to guarantee additional water resources for irrigation.

Recycling agricultural drainage water needs to be promoted nationwide, following the example of a few U.S. states, to avoid water waste and the consequent pollution of lakes and rivers. For instance, Florida farmers cannot pour drainage water from their fields into the Everglades and are forced to have dedicated water reservoirs for excess irrigation water that can be reused for crop irrigation. At the local level, the construction of reservoirs for collecting rainfall has been highlighted to partially solve the water-shortage issue and, especially in case of extreme precipitations (e.g., hurricanes), limit flooding.

Water: A Resource Critical to Food Production and Survival**

Elizabeth A. Bihn, Ph.D.
Senior Extension Associate, Department of Food Science, Cornell University, Geneva, New York, United States

Summary

Fresh water is a limited and valuable natural resource that is crucial to human health and survival. Poor quality water continues to result in human illness and disease despite microbial and chemical quality standards for water used for both drinking and in food production. These standards are both inconsistently formulated and irregularly enforced. Factors limiting water availability and impacting water quality include an expanding global population, climate change, aging infrastructure, and industrial uses. Expanding populations, in particular, will continue to drive an increasing strain on fresh water resources. Water and food production-related problems are complex, long-term, and affect everyone, so solutions will require collaboration across diverse disciplines, in addition to public funding. Currently, water protection and conservation are not priority issues among citizens or policy makers in the United States. Although some scientific research is available to support policy development, there continue to be members of the public, in addition to policy makers, who view scientists and scientific data with suspicion. The lack of science-based policies to protect water combined with changing water availability may negatively impact human health and survival, and food production and safety.

Current realities

Food and water are two of the most essential human needs. For this reason, water availability, water quality, food production, and food safety are global concerns. Although the overall U.S. population and water consumption rates seem fairly steady, fresh water consumption in specific U.S. regions and in other regions of the world is predicted to continue to increase for many reasons, including population shifts and expansion, uncontrolled usage, and climate change. Water quantity is not the only issue: fresh water demand also includes quality requirements to meet intended uses and to avoid human illness and disease.

Although no organization has authority to enforce global water quality standards, the World Health Organization (WHO) provides international leadership in defining parameters and criteria to help achieve water quality goals. In the case of drinking water, WHO guidance states that *"The judgment of safety, or what is an acceptable level of risk in particular circumstances, is a matter in which society as a whole has a role to play. The final judgment as to whether the benefit resulting from the adoption of any of the guidelines and guideline values as national or local standards justifies the cost is for each country to decide."* U.S. drinking water standards, set by the U.S. Environmental Protection Agency (EPA), are comprised of lists of contaminants and their maximum contaminant levels and are designed, in general, to result in lower than a 1 in 10,000 likelihood of adverse effect to those who consume the water.

The need for and use of water are very complex issues. There are human needs for drinking water and food, but also for energy. A current example of water and energy overlap is the U.S. natural-gas boom, driven by controversial high volume hydraulic fracturing (HVHF). These practices in gas exploration involve the use and contamination of large volumes of fresh water such that water quality is degraded. The gas companies support the current costs associated with extraction and disposal but will likely not shoulder long-term costs for necessary regulatory enforcement or impact to the environment. Monitoring and determining the overall impacts to health and the environment are difficult because gas companies do not disclose the chemicals and amounts used in the extraction process, however initial scientific studies outline several reasons for concern and caution. HVHF gas exploration highlights how water issues can compete with other important issues such as energy resource development.

Water quality and availability issues converge during fresh produce production. The use of water for cropland irrigation has a significant impact on stream flow and resulting water availability. The quality of water used in the production of fresh produce is currently the focus of many research programs throughout the world because contaminated water is believed to be the cause of many produce-associated foodborne illnesses and deaths. In the U.S., fresh produce commodity organizations were the first to establish microbial quality criteria for water used in fruit and vegetable production. The Food Safety Modernization Act (FSMA)-proposed Produce Safety Rule has also outlined microbial water standards used in the production of fresh produce. The basis for the irrigation water component of these criteria was a set of EPA recreational water quality standards for activities that result in full body contact with water, such as swimming. Although understandable with regard to intent, the adoption of these criteria to limit health risks in food production lacks a solid scientific foundation for the following reason: the ultimate dose received by a consumer from irrigated produce is fundamentally different from the ultimate dose from full-body exposure to swimming water. Water quality requirements are also a concern because if farmers cannot meet the water quality standards, they may decide to grow crops other than fruits and vegetables. This could result in less consumer access to fresh produce and likely increased prices. In a nation facing heart disease and obesity epidemics, the focus should be on encouraging the production and consumption of fresh produce, especially if there is not sufficient science to prove a human health risk from use of the water.

Water issues are also impacted by the ability of the public and policy makers to understand scientific research results and support science-based policies. As demonstrated by the perceived controversy over the reality of global warming, the public and policy makers in the U.S. often have difficulty evaluating the significance and credibility of scientific reports. The global climate change controversy is a clear example of small groups of individuals ignoring science-based research and effectively using uncertainty to derail effective policy development. A general lack of science literacy in the U.S. public hinders science-based policy development and communication between scientists, the public, and policy makers.

Scientific opportunities and challenges

Fresh water availability, quality, and use need to be intensively studied to determine how current resources can be managed to sustain human populations in the long term. To effectively address opportunities and challenges related to water quality and availability and its involvement in food production, teams should be assembled that include experts from natural and social science disciplines. Current practices and policies should support infrastructure improvements to conserve water during distribution to benefit public citizens, food production facilities, and other businesses dependent on water.

A significant challenge that must be overcome if science-based policies are to be developed is the lack of effective communication between scientists, policy makers, and the public. Trust in scientific research is very low among the public. This is caused by many things, including communication with the public being a very low priority for scientists and research institutions such as universities, as well as a lack of scientific literacy among the public and policy makers.

Policy Issues
- Recognize that clean, potable water is a limited and valuable natural resource critical to human health and survival. Support public funding and coordination of a team of experts with diverse scientific and technical

expertise to study key issues and suggest both research priorities and action plans related to water availability (both quantity and quality of water). This policy goal should interface with the policy goal outlined next.

- Create a national outreach and communication effort to improve the public's scientific literacy related to the issues of water, food production, and water-relevant aspects of climate change. Presentation of complex ideas through simple, clear messaging requires a dedicated, funded, and focused effort. This could be achieved through public service announcements designed for the general public as well as through the development of curriculum modules that can be incorporated in K-12 classrooms. These modules could incorporate core curriculum concepts so they meet state educational requirements while using water, food production, and climate change as the subject. Public universities should enhance their efforts to transparently share how research is conducted and published to improve public understanding and trust. All researchers who receive federal funding should be asked to interface with the public about their research either through a written document or a public webinar. These resources could be catalogued and distributed through public libraries, universities, cooperative extension, public television, or other accessible media outlets. Public scientific literacy *must* be enhanced so that, as a nation, we are better able to utilize the historic U.S. investment in science resources and excellence to address problems related to water and food production through unbiased science rather than economic and political motivations.

- Bolster regulations and enforcement to ensure that use of large volumes of water that result in chemical contamination can either be effectively remediated or that alternative strategies be developed to minimize water use. Companies that use large volumes of water should have to register in a manner similar to the 2002 Bioterrorism Act for food production facilities. This may require personnel in the appropriate government agencies to oversee the process including monitoring and enforcement. Penalties should be severe enough to encourage following developed regulations.

- Develop and support conservation steps that lead to sustainable water use. In the U.S., an easy conservation step is the replacement of aging water distribution systems. Funding water infrastructure improvement through state grants or federal matching grants to municipalities should be considered to distribute and use water resources effectively.

- Establish water quality standards for the production of fresh fruits and vegetables based on science generated in-field production environments. To do this, the government would need to establish an acceptable public health risk standard as a benchmark so that researchers could have a standard for risk comparison. In establishing the standard, the impact it would have on fresh produce production should be considered, since production will impact availability and cost, which in turn impacts consumption and human health.

References

United States Food and Drug Administraton. (2013). Food Safety Modernization Act Proposed Rule for Produce Safety, Docket No. FDA-2011-N-0921.

Gleick, P.H. and M. Palaniappan. (2010). Peak water limits to fresh water withdrawal and use. *PNAS* 107(25):11155-11162.

Milly, P.C.D., K.A., Dunne, and A.V. Vecchia. (2005). Global pattern of trends in streamflow and water availability in a changing climate. *Nature* 438: 347-350.

*** A policy position paper prepared for presentation at the conference on Food Safety, Security, and Defense (FSSD): Focus on Food and Water, convened by the Institute on Science for Global Policy (ISGP) October 20–23, 2013, at the University of Nebraska–Lincoln.*

Debate Summary

The following summary is based on notes recorded by the ISGP staff during the not-for-attribution debate of the policy position paper prepared by Dr. Elizabeth Bihn (see above). Dr. Bihn initiated the debate with a 5-minute statement of her views and then actively engaged the conference participants, including other authors, throughout the remainder of the 90-minute period. This Debate Summary represents the ISGP's best effort to accurately capture the comments offered and questions posed by all participants, as well as those responses made by Dr. Bihn. Given the not-for-attribution format of the debate, the views comprising this summary do not necessarily represent the views of Dr. Bihn, as evidenced by her policy position paper. Rather, it is, and should be read as, an overview of the areas of agreement and disagreement that emerged from all those participating in the critical debate.

Debate conclusions

- Recognizing that water is a compromised resource in terms of quality and availability is essential across governments, industry and the general population. Increasing interactions among the scientific community, policymakers, and the public is necessary to address issues regarding effective management of resources to ensure the availability of quality of water.

- Policy development to incentivize international, national, and local water conservation programs and provide strict enforcement of water regulations are needed to protect global surface and ground water resources.

- Since public trust in science has decreased, in part due to a diminished level of scientific literacy in the general population, it is increasingly critical that the relevance of scientific research to individual lifestyles be effectively communicated. An emphasis on integrating critical thinking skills in school curriculum is a necessary foundation for improving scientific literacy.

- The development of an effective and continuous dialogue between scientists and policymakers is needed to address critical water quality and supply Issues to ensure that the results from scientific research and technological development are properly integrated into policy decisions impacting the management of water systems. Expanding efforts to communicate current and relevant research outputs is needed to strengthen public understanding of how scientific research projects are related to their individual lifestyles.

Current realities

While water and food are essential to the sustainability of our global population, freshwater resources are gradually becoming limited because of the effects of climate change, expanding populations, and growing industrial use. The risk to agricultural production is the increasingly unstable availability and wavering quality of the freshwater supply.

A lack of public trust in credible science is of considerable concern to the scientific community. There was consensus that the scientific community is trustworthy and generally behaves in accordance to ethical practices. However, the lack of scientific literacy in the general public has led to an unwarranted mistrust of the motives of the scientific community in general and the relevance of its research. It was therefore agreed that there is a current and critical need to stress communication, outreach, dialogue, and education to improve scientific literacy among the general public, especially as it pertains to the critical issues concerning water resources.

Scientists have the responsibility to communicate their research findings and the implications of those findings to the public, but often this communication is poor or does not exist. There is therefore a need for a concerted commitment from the scientific community to communicate their research results in ways that are intelligible to those who do not share the same level of scientific expertise.

While there are government agency grants that require communication to the public (e.g., the United Kingdom Department of International Development requires 10% of every research grant to be allocated for communication), government funding does not necessarily mandate that recipients actually communicate their research findings with the public. In the United States, there are policy professionals who actively search for data and results of federally funded research, but find that they cannot easily locate the information or must submit Freedom of Information Act requests to gain access to the information. Limited accessibility to government-funded scientific research can impact public perception of the scientific community's accountability to the public and further perpetuate public distrust of scientific research.

There are different types of water pollution, and in some cases water resources can be cleaned or recycled, but in other cases there are trace amounts (e.g., parts per billion, parts per trillion) of chemicals that can lead to long-term, negative consequences (e.g., increased cancer rates arising from contaminated drinking supplies). Both major and minor pollution of surface and ground water supplies require oversight and strict regulation to maintain the long-term sustainability of the water supply. It was argued that industry has not always been held accountable for damage it has inflicted on the water supply, and that penalties given to companies that fail to comply with water regulations do not match the severity of the damage done to the water supply. The true value of water will be reflected when there are regulations that enforce penalties that are equal to the devaluation caused by polluted water.

Scientific opportunities and challenges

Achievable goals to minimize water loss and increase water quality and conservation must be prioritized (e.g., improvement of aging water distribution infrastructure). A major challenge faced in improving water distribution infrastructure is justifying government prioritization of fund allocation toward new water systems in relation to other infrastructural improvements (e.g., bridges, roads). However, local-level public backing for water infrastructure improvement is attainable because citizens are generally more willing to support mitigation of local infrastructural issues that will decrease their utility costs (e.g., fixing leaks in distribution infrastructure could lower water prices for individuals).

The need for scientists to prioritize communicating their research results and making those findings publicly available was strongly advocated as an approach to building public trust, as well as a method to engage policy makers. The development of an organized, barrier-free venue that would provide accessibility to publicly funded research was proposed. It was also proposed that there is a need for a new generation of scientists who are not only trained in research, but are also explicitly trained in communicating science effectively. This expanded expertise in science communication would build public trust and bridge the existing gap of science literacy between experts and nonexperts. Social scientists could also be effectively utilized in translating scientific conclusions for public understanding and subsequent policy design.

Community outreach requirements could become mandatory in grant proposals to ensure that researchers demonstrate direct links between research outcomes and ultimate community impacts (e.g., water quality improvement). However, it was recognized that mandatory outreach would compete with the allocation of funding to primary research. Diverting primary research funding to community outreach is unpopular with some researchers who would rather not prioritize communication over research operations. It was, however, agreed that an essential step in building public trust is providing public accessibility to research results.

It was argued that a project-to-project or grant-to-grant basis is not always the most effective forum for communicating science, as research is not always applicable and germane to current policymaking issues. Rarely does one project produce a complete body of evidence that is adequate for definitive decisions to be made. The challenge remains to mandate science researchers to demonstrate a direct link from science research output to public impact (e.g., proposed legislation or behavioral change).

Improving scientific literacy through assessment of existing school curricula offers an opportunity to address public trust in the scientific community. Educational science materials, approved by experts to ensure that the scientific information is factual, need to be developed and made available to public schools. A heightened emphasis on training students in critical thinking skills is also needed. Improved critical thinking can result in a higher baseline of scientific understanding among the public, which would potentially raise confidence in the conclusions drawn by experts and, perhaps indirectly, influence policy-making through increased scientifically sound communication between constituents and their representatives.

Public distrust in science can stem from alarmist narratives developed for various purposes (e.g., selling exciting news stories). However, once these narratives are established, it is difficult to change public perceptions with rational, science-based arguments. Therefore, scientists need to develop the ability to generate narratives based on credible science of important issues as part of the goal to enhance communication and public understanding.

Policy issues

It is imperative to communicate the results and implications of federally funded research with the public and with policy makers to encourage the development of science-based policies. The questions remains as to who should be responsible for the communication. It was recommended that this could fall within the jurisdiction of the funding agency. Under this scenario, scientists report their research results to the funding agency, which would then synthesize the results and effectively communicate the results and potential policy implications to the public. It was also noted that while communicating outcomes of science research is necessary, there is also the risk of an information overload that could lead to public disinterest.

Policies need to be developed to increase early education on water conservation and other science topics. From an early age, children need increased exposure to science, as well as education on the importance of the role of science, which will prepare students with a widened perspective and equip them with the ability to use and cite scientific reasoning. More universal educational standards (e.g., the Common Core Standards have been designed to tailor curriculum development to current, relevant knowledge that prepare students with the skills needed to be successful in higher education and their careers) need to be implemented by governments to guarantee that students become science literate.

There is zero-risk tolerance for food safety, but water safety standards and new regulations (e.g., United States Department of Agriculture [USDA] policies on water for food-raising and food processing) need to be based on credible science. In addition, separate health standards for different water uses (e.g., drinking water versus agricultural irrigation) need to be developed. The development of water use-specific safety standards can positively impact human health by giving the public confidence in food supply inputs, thus encouraging people to eat more fresh produce.

Conservation steps must be developed and supported by community and private-sector stakeholders, politicians and science researchers. It is also necessary to engage representatives from industries that require heavy water use (e.g., hydraulic fracturing) in the water conservation dialogue. Industry must be held accountable for damage inflicted on the water supply through penalties that are equal to the devaluation of polluted water, especially if the water is polluted in a way that cannot be restored to near its original purity standards. Regulation enforcement can increase the value of water resources.

Innovation and Policy against Hunger in a Water-Constrained World**

Konstantinos Giannakas, Ph.D.
Professor and Director, Center for Agricultural & Food Industrial Organization,
Department of Agricultural Economics, University of Nebraska–Lincoln,
Lincoln, Nebraska, United States

Summary

This paper discusses the role of innovation and policy in the fight against hunger in a water-constrained world. The innovations considered here are genetic modification technologies that combine the provision of agronomic benefits to producers with enhanced nutritional value to consumers (i.e., technologies that combine/stack input and output traits). The development of such technologies requires significant resources and is quite often accompanied by the granting of intellectual property rights (IPRs) to the innovator(s) involved.

The granting of IPRs aims to bolster incentives for research and development (R&D) by providing the innovator with monopoly rights over the new technology. While these rights do increase innovation activity, they can have a significant impact on the price of the new technology and, through this, on the public's access to the innovation.

Understanding that *hunger can be reduced through access to increased quantities of nutritious food offered at affordable prices*, most of the discussion focuses on the effects of different technologies and IPR policies on quantities produced, the quality of production, the prices of food products, and the number of people with access to food in hunger-stricken less-developed countries (LDCs). Once the scientific opportunities and challenges have been identified, the paper highlights some key policy issues shaping the effectiveness of genetic modification innovations and IPR policies in combating hunger around the world.

Current realities

The introduction of genetically modified organisms (GMOs) into the food system and the assignment of IPRs for plant genetic resources are among the most notable features of the increasingly industrialized agrifood marketing system of numerous countries, both developed and developing, around the world. IPRs have provided innovating firms with incentives to aggressively pursue improvements of crop characteristics (such as herbicide tolerance, insect and virus resistance, drought tolerance, and increased nutritional value) through gene splicing techniques, and the agronomic benefits of the genetically modified (GM) products have resulted in their embrace by a significant number of agricultural producers around the world.

In particular, 16 years after their initial commercialization in 1996, GM crops were grown on 170 million hectares worldwide with (i) more than half (52%) of those being planted in developing countries such as Brazil, Argentina, India, China, and South Africa; and (ii) a quarter being planted with biotech crops having multiple (i.e., stacked) traits. Based on James (2013), 17 million farmers in 28 countries grew GM soybeans (47% of global biotech area), maize (27%), cotton (14%), and canola (5%) in 2012. GM papaya, alfalfa, squash, rice, and sugar beet were also cultivated on much smaller areas. The market value of biotech crops in 2012 was $14.8 billion, representing 23% of the global crop protection market and 35% of the global commercial seed market.

Intriguingly, in the midst of this so-called gene revolution, about 1 billion people worldwide are facing malnutrition and hunger, with the majority of these people living in water-constrained regions of Africa and Asia. With GMOs and IPR being at the epicenter of innovation activity in the agrifood system, the question that naturally arises is: *Can GMOs and IPRs help reduce hunger in a water-constrained world?* This paper argues that they can.

Scientific opportunities and challenges

Scientific research on the effects of genetic modification technologies has focused on the effects of different types of GMOs (e.g., first-generation producer-oriented GMOs, second-generation consumer-oriented GMOs, and, lately, GMOs having stacked both input and output traits) on quantities produced, the quality of production (with the output traits of second-generation GMOs being quality-enhancing), the prices of food products, and the number of people that have access to food. Different regulatory and labeling regimes have been considered within this framework.

The research has identified the potential for significant benefits from the development and adoption of appropriate genetic modification technologies for all participants in the agrifood marketing system (Qaim, 2009). An important message of this literature is that *properly designed genetic modification technologies* (i.e., technologies adapted to the idiosyncrasies and needs of an area) can facilitate production, increase yields, reduce production costs, and enhance the nutritional value of food products.

Key input traits of the GMOs needed in the fight against hunger are drought resistance and/or water-use efficiency of plants, as water has been a key constraining factor in many hunger-stricken countries. The necessary output traits (e.g., vitamin, iron, or zinc enhancements), will have to be case-specific and dependent on the nutritional needs of different areas.

Important determinants of the effectiveness in combating hunger of these genetic modification technologies are (i) the public attitudes towards GMOs; (ii) the magnitude and distribution of benefits of the GM technology; (iii) the regulatory and labeling regimes governing GMOs (domestically and internationally); (iv) the structure of the agrifood marketing system; (v) the market power of innovating companies; and (vi) the strength and enforcement of IPRs in LDCs.

While GM technologies *can* result in increased quantities of nutritious food in hunger-stricken LDCs, there are some major challenges in the quest to utilize such technologies in the fight against hunger. These challenges include (i) the limited availability of suitable GM crops and technologies; (ii) the limited capacity for R&D in most LDCs; (iii) the role of NGOs in shaping public attitudes towards GMOs; (iv) the trade relationships of LDCs with countries hostile to GMOs; and (v) the inefficiency of the regulatory system in most LDCs.

Research in the area of IPR enforcement has focused on the effects of different IPR enforcement policies and strategies on the prices and adoption of new technologies, the level of output produced, and the number of people with access to the relevant innovation(s). Different objectives of innovators and governments involved have been considered within this framework.

The level of IPR enforcement has been shown to affect the welfare of the interest groups involved (i.e., producers, consumers, and innovators), and has important ramifications for the pricing and adoption of the new technology. Specifically, the weaker the enforcement of IPRs in a country, the lower the price of the new technology, the greater the technology adoption by producers, and the more consumers who have access to this technology. Since the price of the new technology is inversely related to the level of IPR enforcement, lax IPR enforcement also increases the international competitiveness of domestic producers utilizing this technology (by placing producers in countries where IPRs are more effectively enforced at a cost disadvantage). While lax enforcement of IPRs can benefit *all* biotechnology users in an LDC, it reduces the innovators' rents by diminishing their ability to obtain value for their biotech traits (Giannakas, 2002).

Since weak IPR enforcement benefits the LDC producers and consumers while reducing the rents accruing to the innovator(s), the level of enforcement in the LDC will be determined by the political preferences of the government and the weight it places on innovator rents. The less importance the domestic government places on (usually foreign) innovator rents, the lower the level of IPR protection. Factors affecting the importance the domestic government places on innovator rents (and, thus, its enforcement policy) include (i) the political influence of the innovating firm in the LDC; (ii) the bilateral relationship with the country of origin of the innovating firm; (iii) the severity of the sanctions in the case that the LDC is successfully held to have imperfectly enforced the innovator's IPRs; (iv) the conjectures of the domestic government regarding the effect of its enforcement policy on the future development of (and domestic access to) new technologies; (v) the role of NGOs in shaping IPR policies (e.g., by challenging patents and lobbying for certain provisions); and (vi) the size of the enforcement costs.

It is important to note that, while the above discussion (and most of the relevant literature on the topic) assumes that innovators desire the strong enforcement of their IPRs, there might be cases that the innovating firms find it optimal to not enforce their IPRs in hunger-stricken LDCs. In fact, there could be cases that innovators find it optimal to provide free access to their new technology in these countries. For instance, if this enforcement strategy increased the innovator's goodwill in the LDCs (that get the technology for free) but also in developed countries (that can now associate the innovator *and* the innovation with a noble humanitarian endeavor), the benefits to the innovator could easily outweigh the lost royalty fees from these LDCs. If done correctly, such an IPR strategy could result in significant benefits for hunger-stricken LDCs, the innovating firms, *and* the image of (and public attitudes towards) agricultural biotechnology as a whole. The latter could be particularly important in places like the European Union where the consumer opposition to GMOs has shaped the regulatory response to these organisms with significant ramifications for many hunger-stricken LDCs trading with the E.U.

Policy issues

Since GM technologies can play a significant role in the fight against hunger, it is important to consider the key issues affecting the development of appropriate genetic modification technologies, and the adoption of such technologies in hunger-stricken LDCs.

- Stacked genetic modification technologies, which combine drought resistance and water-use efficiency of plants with quality enhancing attributes relevant to the nutritional needs of the local population, need to be developed. This development must be done by innovating firms and universities with relevant expertise, in collaboration with local experts. If possible, this work should be done domestically in the countries of need and/or at appropriate facilities abroad, with funding for research, capacity building, and training of local scientists provided by government programs (e.g., U.S. Agency for International Development, National Science Foundation, U.S. Department of Agriculture), the World Bank, major foundations and philanthropists, and NGOs.

- The adoption of new technologies will require a change in public attitudes towards GMOs. This can be achieved by local universities and research stations, government agencies, and NGOs providing information about the potential benefits of relevant GM crops.
- Enhancing technology adoption and consumer access to new food products can also be facilitated by reduced prices for these products. In hunger-stricken LDCs, this can be achieved by lax or no enforcement of IPRs, coupled with R&D subsidies and/or public R&D.
- Efficient regulatory systems are needed to evaluate new innovations in a timely manner.

References

James, C. (2013). Global Status of Commercialized Biotech/GM Crops: 2012. ISAAA Brief No. 44, International Service for the Acquisition of Agri-biotech Applications, Ithaca, N.Y.

Qaim, M. (2009). The economics of genetically modified crops. *Annual Review of Resource Economics*, 1, 665-694.

Giannakas, K. (2002). Infringement of intellectual property rights: Causes and consequences. American Journal of Agricultural Economics, 84, 482-494.

** A policy position paper prepared for presentation at the conference on Food Safety, Security, and Defense (FSSD): Focus on Food and Water, convened by the Institute on Science for Global Policy (ISGP) October 20–23, 2013, at the University of Nebraska–Lincoln.

FOCUS ON FOOD AND WATER 83

Debate Summary

The following summary is based on notes recorded by the ISGP staff during the not-for-attribution debate of the policy position paper prepared by Dr. Konstantinos Giannakas (see above). Dr. Giannakas initiated the debate with a 5-minute statement of his views and then actively engaged the conference participants, including other authors, throughout the remainder of the 90-minute period. This Debate Summary represents the ISGP's best effort to accurately capture the comments offered and questions posed by all participants, as well as those responses made by Dr. Giannakas. Given the not-for-attribution format of the debate, the views comprising this summary do not necessarily represent the views of Dr. Giannakas, as evidenced by his policy position paper. Rather, it is, and should be read as, an overview of the areas of agreement and disagreement that emerged from all those participating in the critical debate.

Debate conclusions

- The introduction of GMOs has been a key component in improving agricultural production in an increasingly industrialized and global food system. If properly designed, including its integration by producers around the world, such GMO agricultural technologies can benefit the global population by increasing production yields, reducing agricultural costs, and enhancing the nutritional value of food products.

- Because the public perception of GMOs determines the adaptability and integration of genetic modification technologies to communities, biotechnology innovation companies need to employ subject matter experts, including academics, from diverse areas of expertise to lead communication and outreach programs designed to increase public trust in the biotechnology industry.

- Since IPRs have provided incentives for agricultural innovation that has led to significant crop improvements, enforcement of IPRs is necessary to foster agricultural innovations. However, some private companies have chosen to offer royalty-free licenses to producers in less-affluent countries to provide access to agricultural benefits in parts of the world where it is needed the most. This goodwill of the innovating firms creates value from market growth and other positive externalities from the implementation of tailored technologies, and therefore needs to be encouraged.

- To foster effective communication and collaboration on research and development opportunities tailored to the distinctive characteristics of specific water-constrained areas, innovating firms must develop partnerships with local scientists from communities in which agricultural technologies can significantly improve the food supply. New programs are also needed for collecting local scientific and cultural information as well as the most effective sources of credible information to be used as developmental data in innovation research.

Current realities

There was consensus that GMOs and IPRs play a significant role in the global fight against hunger. Recent research has shown that GMOs can provide global benefits by increasing agricultural production in water-constrained areas. The development of stacked technologies (i.e., genetically modified organisms designed to have multiple beneficial traits) can benefit producers as well as consumers (e.g., drought resistance traits will benefit farmers and enhanced vitamin content traits will benefit consumers), especially in hunger-stricken countries that lack reliable water supplies.

Biotechnology is one in a large range of technologies (e.g., fertilizers, irrigation) that are needed to improve the stability of the global food supply in a water-constrained world. However, not all countries are proponents of the increasing role of biotechnology. The European Union, the largest player in global agricultural, has the most stringent GMO regulations in the world. E.U. purity standards require labeling on any food product containing more than 0.9% of approved GMOs (most approved GMOs are strictly used for animal feed).

There is a lack of public understanding of the process that innovating firms are required to undertake to introduce new technologies to the market as well as the amount of money spent by companies to gain regulatory approval for genetically modified crops. A misconception exists that governments uninhibitedly allow new technologies into the marketplace and that consumers are used to test the new products. There also exists a perception of arrogance that is associated with biotechnology that hurts the industry and limits the opportunities in achieving wider acceptance of GMOs. Biotechnology in the field of medicine, however, is generally accepted by the public because people more easily recognize the specific health benefits (i.e., therapies to illness or ailments). With any new technology, there is almost always initial consumer distrust.

Less-affluent countries that suffer from malnutrition and hunger, notably in water-constrained areas of Asia and Africa, often lack a regulatory framework for GMO adoption. Cultural barriers also exist that inhibit the adoption of GMO technologies (e.g., some farmers are reluctant to use unfamiliar new seeds).

It was noted that most of current literature published on GMOs and IPRs perpetuates the assumption that the innovating companies are maximizing profits. As a result, this literature uniformly seeks strong global enforcement of IPRs. However, recent research has revealed situations in which it is optimal for the innovators to withhold enforcement of IPRs in less-affluent countries. For example, weak enforcement of IPRs in less-affluent countries lowers the price of a new technology and brings a greater level of technology adoption, which results in greater consumer accessibility to the benefits of the innovation.

Research has identified specific conditions under which innovating firms could find it optimal to make new technologies available for free. Agricultural biotechnology designed to increase food production yields and/or decrease the use of water can be beneficial in water-constrained areas that suffer from hunger. However, less-affluent countries often cannot afford to pay to access these technologies. Some innovating firms have granted royalty-free licenses to these countries with the intention of offering help in the fight against hunger in areas of the world that are the most greatly in need. Research has shown that profitability can be maintained by innovators because the losses in terms of royalty fees from less-affluent countries will be offset in terms of goodwill and continued support elsewhere in the world.

Producers in countries with weak IPR enforcement have a competitive advantage over the producers in countries that strictly enforce remuneration for intellectual property licensing. Some countries have found it beneficial to willingly accept prosecutions from the World Trade Organization (WTO) for IPR violations because the net gain from increased agricultural production (with unlicensed use of innovation) remained higher with penalty payouts than it would have been had they complied with global trading rules and paid licensing and royalty fees (e.g., Argentina has had historically weak enforcement of IPRs, where these adverse incentives have resulted in a net benefit). There is a need to address the severity of WTO sanctions given the counterproductive nature of the existing penalty structure.

IPR policy is perhaps determined by how much influence an innovating firm has in the country, and how strong the bilateral relationship of country of origin of the innovating firm is with the country adopting the innovation. The politics of the adopting country can influence the enforcement of IPRs, such that if an administration cares about the impact of its behavior on the future access of producers to technologies, then the enforcement approach might be more rigorous, despite the possibility of suffering domestic political disapproval. Conversely, if an administration has shorter election cycles, strong enforcement might not be a priority.

Scientific opportunities and challenges

Although GMOs have the potential to increase the quantity and quality of food production, a major challenge is the limited number of suitable technologies tailored to fit the needs of less-affluent countries. These countries generally have a limited domestic capacity for research and development, and have trade relationships with countries that are hostile to GMOs (e.g., the E.U.).

A significant barrier to the widespread adoption of agricultural innovations is consumer aversion to GMOs, which is most notably present in the E.U. It was argued that the strict regulatory approach to GMOs by the E.U. indirectly places strain on less-affluent countries. Not only does the E.U. aversion to GMOs subject the countries they trade with to an economic disadvantage (e.g., by imposing trade barriers on GMO crops that are cheaper to produce than non-GM crops), but also creates potential health disadvantage (i.e., malnutrition and hunger due to lower potential food production) as well. Trade barriers affect the regulatory response to GM agricultural production and less-affluent countries generally have weak or nonexistent internal regulatory and approval systems.

It is important to improve communication supporting the continuing development of biotechnological agricultural innovations as well as addressing public trust. Better transmission of information can build public trust. Communicating benefits of utilizing GMO technology, coupled with the existing and potential risks of GMO integration, can change the public attitude towards GMOs and biotechnology. Communications about GMO innovations can include altruistic biotechnology endeavors (e.g., the Golden Rice Project, companies granting royalty-free licensing to less-developed countries) to gain consumer trust. However, it can be challenging for innovating firms to communicate their research and innovations without sounding as if they are purely motivated to enhance their public image.

Local scientists can assess the potential benefits of creating site-specific GMO technologies for communities burdened by agricultural problems (e.g., unpredictable water supply) or human health challenges (e.g., vitamin deficiency), and present the potential options and opportunities to these communities. Less-affluent countries that cannot afford the licensing fees of their site-specific technology options will need to be provided with these technologies by the innovating firms at no or low costs. If marked growth and other positive externalities are realized by the implementation of tailored technologies, the goodwill of the innovating firms will not go unrecognized by the international community, and will thus create value in the companies.

Policy issues

The next generation of policymakers and citizens are likely to inherit the ongoing issue of building public acceptance of GMOs. It was suggested that public policy is influenced by the experience of the policymaker in the formative years, and that the E.U. aversion to GMOs is partly due to the current cohort of E.U. policymakers having grown up in an environment of food abundance. Therefore, it is challenging to convince E.U. policymakers that new technologies are needed to produce more food. Public perception and perceived benefits are predicted to shift as a new generation will influence decision makers to consider the instability of food supply in less-developed countries.

Because innovators need to be compensated for their research and development investments, it is necessary to demand IPR enforcement in most countries. There is a need to develop strategies to protect the intellectual property interest (i.e., innovator rents) while allowing a significant amount of "free ridership" for less-affluent countries. It remains to be determined how producers and consumers of more-affluent countries will value this altruism, as well as whether these countries would be willing to pay more for licensing, given that licenses will be provided for free in less-affluent countries. More-affluent countries are more likely to participate in funding agricultural innovation if provided with certainty that the funding will result in benefit for the populations in less-affluent countries (e.g., fewer people suffering from hunger). Innovating firms must effectively communicate outcomes of their charitable efforts to more-affluent countries to increase willingness to pay.

Innovating firms have the opportunity to take the lead in educating the public regarding the benefits of emerging agricultural technologies and must select the most credible professionals to lead the communications. Research has found that academics (i.e., university professors) are the professional group that consumers find most trustworthy, and therefore must be utilized in communicating genetic modification innovations. Utilizing an interdisciplinary academic team (e.g., scientists, social scientists, economists) to address communication would result in increased credibility by the public of the GMO industry.

Innovating firms and universities in more-affluent countries need to develop and maintain associations with scientists from local communities that are in need of agricultural innovation to combat hunger and agricultural challenges. In general, local scientists and food producers are most familiar with local needs (e.g., areas affected by water constraints). Collaboration among public institutions, universities, and innovating firms is needed for a coordinated effort in research and development. If the technology and infrastructure are not available in the countries needing innovation, then research and development must be done abroad, but in close communication with the countries for which the work is being tailored.

Global economic development can be achieved by funding biotechnology research and integration. New programs are needed for collecting local scientific and cultural information to be used as developmental data for innovation research. These programs would seek the most effective sources of credible information in less-affluent countries for the purpose of communicating the benefits of agricultural biotechnologies. These programs may be funded through multiple sources (e.g., government agencies, private sector, philanthropic organizations, and nongovernmental organizations).

Acknowledgment

Numerous individuals and organizations have made important contributions to the Institute on Science for Global Policy (ISGP) program on Food Safety, Security, and Defense (FSSD). Some of these contributions directly supported the efforts needed to organize the invitation-only ISGP conference, *Focus on Food and Water,* convened in cooperation with the University of Nebraska–Lincoln at the Embassy Suites in Lincoln, Nebraska, October 20–23, 2013. Other contributions aided the ISGP in preparing the material presented in this book, including the eight invited policy position papers and the summary record, without attribution, of the views presented in the discussions, critical debates, and caucuses that ensued.

The ISGP greatly appreciates the willingness of those in the scientific and policy communities to be interviewed by the ISGP staff, who organized the content of this ISGP conference. Of special significance were the efforts of those invited by the ISGP to present their views of the scientific and societal impact of genomics through their policy position papers. Their willingness to engage policy makers and other scientists in the vigorous debates and caucuses that comprise all ISGP conferences was especially appreciated. The biographies of these eight authors are provided in this ISGP book.

The success of every ISGP conference critically depends on the active engagement of all invited participants in the often-intense debates and caucuses. The exchange of strongly held views, innovative proposals, and critiques generated from questions and debates fosters an unusual, even unique, environment focused on clarifying understanding for the nonspecialist. These debates and caucuses address specific questions related to formulating and implementing effective public and private sector policies. The ISGP is greatly indebted to all those who participated in these not-for-attribution debates and caucuses.

The members of the ISGP Board of Directors also deserves recognition for their time and efforts in helping to create a vital, increasingly relevant not-for-profit organization is addressing many of the most important societal questions of our time. Their brief biographical backgrounds are presented at the end of this book.

The energetic, highly professional work of the ISGP staff merits special acknowledgment. The staff's outstanding interviewing, organizing, and writing skills remain essential to not only organizing the conference itself, but also to recording the often-diverse views and perspectives expressed in the critical debates, capturing the areas of consensus and actionable next steps from the caucuses, and persevering through the extensive editing process needed to assure the accuracy of the material published here. All of the staff members' work is gratefully acknowledged. Their biographies are provided in this book.

ISGP programs are financially supported by government agencies and departments and through gifts from private-sector entities and philanthropic individuals. Specifically, the ISGP conference on *FSSD: Focus on Food and Water*, received funding for its general activities from the U.S. Department of State and the U.S. Department of Health and Human Services. The ISGP also benefited greatly from generous gifts provided by the MARS Corp., GlaxoSmithKline, Cargill Inc., and Edward and Jill Bessey.

> Dr. George H. Atkinson
> Founder and Executive Director
> Institute on Science for Global Policy
> December 12, 2013

ISGP books from ISGP conferences listed below are available to the public and can be downloaded from the ISGP Web site: www.scienceforglobalpolicy.org. Hardcopies of these books are available by contacting Jennifer Boice at jboice@scienceforglobalpolicy.org.

ISGP conferences on, or related to, Emerging and Persistent Infectious Diseases (EPID):

- *EPID: Focus on Antimicrobial Resistance,* convened March 19–22, 2013, in Houston, Texas, U.S., in partnership with the Baylor College of Medicine
- *21st Century Borders/Synthetic Biology: Focus on Responsibility and Governance,* convened December 4–7, 2012, in Tucson, Arizona, U.S., in partnership with the University of Arizona.
- *EPID: Focus on Societal and Economic Context,* convened July 8–11, 2012, in Fairfax, Virginia, U.S., in partnership with George Mason University
- *EPID: Focus on Mitigation,* convened October 23–26, 2011, in Edinburgh, Scotland, U.K., in partnership with the University of Edinburgh.
- *EPID: Focus on Prevention,* convened June 5–8, 2011, in San Diego, California, U.S.
- *EPID: Focus on Surveillance,* convened October 17–20, 2010, in Warrenton, Virginia, U.S.
- *EPID: Global Perspectives,* convened December 6–9, 2009, in Tucson, Arizona, U.S., in partnership with the University of Arizona.

ISGP conferences on Food Safety, Security, and Defense (FSSD):

- *FSSD: Focus on Food and Water,* convened October 20–23, 2013 in Lincoln, Nebraska, U.S., in partnership with the University of Nebraska–Lincoln.
- *FSSD: Focus on Innovations and Technologies,* convened April 14–17, 2013, in Verona, Italy.
- *FSSD: Global Perspectives,* convened October 24, 2012, in Arlington, Virginia, U.S., in partnership with George Mason University.

Biographical information of Scientific Presenters and Keynote Speakers

Scientific Presenters

Prof. Roberto Lenton, Ph.D.
Prof. Roberto Lenton is Founding Executive Director of the Robert B. Daugherty Water for Food Institute at the University of Nebraska, and Professor of Biological Systems Engineering at the University of Nebraska–Lincoln. He has previously served as Chair of the World Bank's Inspection Panel, Senior Adviser at Columbia University's Earth Institute, Director of the Sustainable Energy and Environment Division of the United Nations Development Program in New York, Director General of the International Water Management Institute in Sri Lanka (an organization he was instrumental in establishing), and as Program Officer in the Rural Poverty and Resources Program of the Ford Foundation in New Delhi. Additionally, he has held academic positions at Columbia and the Massachusetts Institute of Technology. Prof. Lenton is a respected expert on issues related to water management, food security, sustainable agriculture, and responsible use of resources, and has co-authored and edited numerous books on the subject, including "Applied Water Resources Systems" and "Integrated Water Resources Management in Practice." He served as Chair of the Water Supply and Sanitation Collaboration Council and the Technical Committee of the Global Water Partnership, and co-Chair of the Millennium Project Task Force on Water and Sanitation.

Dr. Elizabeth Bihn, Ph.D.
Dr. Elizabeth Bihn is a Senior Extension Associate in the Department of Food Science at Cornell University. She is currently the Director of the Produce Safety Alliance, a collaborative project between Cornell University, the United States Department of Agriculture, and the Food and Drug Administration that seeks to increase understanding of the principles of Good Agricultural Practices and to facilitate the implementation of food safety practices on fresh fruit and vegetable farms and in packinghouses. Dr. Bihn is also the program coordinator for the National Good Agricultural Practices Program, based at Cornell. The main focus of her work is to reduce microbial risks to fresh fruits and vegetables through research and extension

programs developed for and in collaboration with growers, farm workers, produce industry personnel, students, teachers, and consumers.

Dr. Iain Wright, Ph.D.

Dr. Iain Wright is a Director of the International Livestock Research Institute (ILRI), one of centers of the Consultative Group on International Agricultural Research, and headquartered in Nairobi, Kenya. He is currently the Program Leader for Animal Science for Sustainable Productivity, a $15 million global program that aims to increase productivity of livestock systems in developing countries through high quality animal science (breeding, nutrition, and animal health) and livestock systems research. He is based in Addis Ababa, Ethiopia, where he is also the ILRI Director General's Representative in Ethiopia and Head of the Addis Ababa campus, where more than 300 staff members are located. From 2006 to 2011, he was the ILRI Regional Representative for Asia, coordinating ILRI's activities in South Asia, South East Asia, and China, working with a wide range of public, private, and civil organizations. Prior to joining ILRI, Dr. Wright held a number of posts at the Macaulay Institute, Scotland, including managing the Institute's research program on agroecosystems, which included research on land-use change and land-use systems and their impacts. From 2003 to 2006, he was Chief Executive of Macaulay Research Consultancy Services Ltd., the Macaulay Institute's knowledge transfer company

Prof. Alvin Smucker, Ph.D.

Prof. Alvin Smucker is Professor of Soil Biophysics at Michigan State University (MSU). Additionally, he is Director of the MSU Subsurface Water Retention Technology Program, Visiting Chair Professor of Soil Science, Scottish Agriculture College, and Visiting Research Soil Scientist, Argonne National Laboratories. His primary research interests include plant root development, biogeochemistry and biophysics of soil carbon sequestration, and soil water retention and quality surrounding Red Cedar River on the MSU campus (MSU-WATER) and the Great Lakes (Great Lakes Commission). His soil biophysical and plant root research activities have resulted in more than 335 peer-reviewed journal articles, book chapters, conference proceedings, abstracts, and research reports. Prof. Smucker has received many awards and honors, including a 2005 Distinguished Faculty Award at Michigan State University, two separate Alexander von Humboldt Research Awards (2003 and 2010), and a 2000 Senior Faculty Research Award from the MSU Chapter of Sigma Xi Research Society. He is a Fellow of five national and international scientific societies, including the American Association for the Advancement of Science.

Dr. Robert E. Brackett, Ph.D., M.S.
Dr. Robert Brackett is Vice President of the Illinois Institute of Technology (IIT) and Director of the Institute for Food Safety and Health (IFSH) at IIT, and Co-Director of the National Center for Food Safety and Technology (NCFST). Dr. Brackett has nearly 30 years of experience in scientific research in industry, government, and academia. His work has focused primarily on the areas of food safety, defense, and nutrition. Prior to his post at IFSH, he served as Senior Vice President and Chief Science and Regulatory Officer for the Grocery Manufacturers Association (2007 to 2010); was Director of the U.S. Food and Drug Administration's (FDA) Center for Food Safety and Applied Nutrition, and held professorial positions within North Carolina State University (Raleigh) and the University of Georgia. Dr. Brackett is a fellow of the International Association for Food Protection and American Academy of Microbiologists, and a member of the International Association for Food Protection, Institute of Food Technologists, and the American Society for Microbiology. He has been honored with the FDA Award of Merit, the International Association for Food Protection's President's Appreciation Award, and the William C. Frazier Food Microbiology Award.

Ms. Debbie Reed, M.Sc.
Ms. Debbie Reed is Executive Director of the Coalition on Agricultural Greenhouse Gases, a multistakeholder coalition supporting policies and programs to incentivize voluntary agricultural sector mitigation of greenhouse gases. Ms. Reed is also the Policy Director for the International Biochar Initiative, where she was Executive Director for the past six years. She is also the President and Director of DRD Associates, providing strategic and policy support for national and international agricultural mitigation strategies for global climate change. Previously, Ms. Reed worked for President Bill Clinton at the White House Council on Environmental Quality as the Director of Legislative Affairs and Agricultural Policy for the White House Climate Change Task Force. She was also a Senior Legislative Assistant for U.S. Sen. J. Robert Kerrey (D-NE), where she handled environmental, natural resource/agriculture, and energy issues. In previous positions at the U.S. Department of Agriculture and at several public health-oriented institutions, Ms. Reed's work focused on federal agricultural, food safety, and human nutrition policy.

Dr. Rob Atwill D.V.M., M.P.V.M., Ph.D.
Dr. Rob Atwill is Director of the Western Institute for Food Safety and Security at the University of California (UC), Davis. Additionally, he has been part of the faculty in the School of Veterinary Medicine at UC Davis since 1994. His research and extension program has focused on key processes governing the fate, transport,

and dissemination of zoonotic diseases and their role in microbial water quality and food safety. He conducts epidemiological studies on the occurrence of waterborne zoonotic pathogens in rural and agricultural ecosystems located throughout California, across the United States, and internationally. Through this research, Dr. Atwill and research partners have helped pioneer the theory and application of on-farm vegetative buffers for minimizing the ability of microbial pathogens to become a waterborne hazard to humans and animals. These activities have produced more than 100 peer-reviewed publications and more than 150 lay articles and abstracts. Additionally he has conducted more than 180 presentations, workshops, and symposia in his area of research.

Dr. Konstantinos Giannakas, Ph.D.

Dr. Konstantinos Giannakas is Professor of Agricultural Economics at the University of Nebraska–Lincoln (UNL). He is also Director of the Center for Agricultural and Food Industrial Organization (CAFIO) at UNL, and Program Director of the U.S. Department of Agriculture-funded CAFIO Policy Research Group. Dr. Giannakas has received numerous awards for his teaching and research, and has been an expert consultant for the U.N. Food and Agriculture Organization. Prior to joining the faculty at UNL, Dr. Giannakas was a lecturer and research scientist at the University of Saskatchewan, Canada, and has held visiting appointments at the Mediterranean Agronomic Institute of Chania, Greece, Wageningen University, The Netherlands, and the University of Agricultural Sciences, Vienna, Austria. He is an associate editor for the *Journal of Agricultural & Food Industrial Organization*, and previously an associate editor for the *American Journal of Agricultural Economics*.

Keynote Speakers

Dr. Ken Cassman, Ph.D.

Dr. Ken Cassman is the Robert B. Daugherty Professor of Agronomy at the University of Nebraska–Lincoln. He is also the Chair of the Independent Science and Partnership Council of the Consultative Group on International Agricultural Research. He was Chair of the Department of Agronomy and Horticulture at the University of Nebraska from 1996–2004. Dr. Cassman's research and teaching have focused on ensuring local and global food security while conserving natural resources and protecting environmental quality for future generations. He served on agricultural development projects in Brazil and Egypt (1980–1984), was on the faculty at University of California, Davis from 1984–1990, and served as Head of Agronomy, Plant Physiology, and Agroecology at the International Rice Research Institute in the Philippines (1991–1995). He is best known for his publications on crop yield potential and yield gap analysis, nitrogen use efficiency, and global food security, and as co-author on the textbook, "Crop Ecology." Dr. Cassman has received a number of professional awards for his contributions in research and education, most recently the 2012 President's Award from the Crop Science Society of America. He is a Fellow of the American Association for the Advancement of Science, the American Society of Agronomy, the Crop Science Society of America, and the Soil Science Society of America. He currently serves as Editor-in-Chief of the journal *Global Food Security*.

Dr. Andrew Benson, Ph.D.
Dr. Andrew Benson is the W. W. Marshall Professor of Biotechnology and the Director of Core of Applied Genomics and Ecology ath the University of Nebraska–Lincoln. His current research focuses on the study of gut microbiome, and specifically, host genetic factors that shape the establishment and composition of this complex ecosystem. His work incorporates next-generation sequencing technologies as a set of high-throughput methodologies for measuring species composition of gut microorganisms. Dr. Benson collaborates with quantitative geneticists, genomicists, bioinformaticians and statisticians to integrate the gut microbiome as a quantitative trait in large-scale genetic association studies in animal models with the ultimate goal of understanding the genetic architecture underlying predispositions to complex lifestyle diseases.

Biographical information of ISGP Board of Directors

Dr. George Atkinson, Chairman
Dr. George Atkinson founded the Institute on Science for Global Policy (ISGP) and is an Emeritus Professor of Chemistry, Biochemistry, and Optical Science at the University of Arizona. He is former head of the Department of Chemistry at the University of Arizona, the founder of a laser sensor company serving the semiconductor industry, and Science and Technology Adviser (STAS) to U.S. Secretaries of State Colin Powell and Condoleezza Rice. He launched the ISGP in 2008 as a new type of international forum in which credible experts provide governmental and societal leaders with the objective understanding of the science and technology that can be reasonably anticipated to help shape the increasingly global societies of the 21st century. Dr. Atkinson has received National Science Foundation and National Institutes of Health graduate fellowships, a National Academy of Sciences Post Doctoral Fellowship, a Senior Fulbright Award, the SERC Award (U.K.), the Senior Alexander von Humboldt Award (Germany), a Lady Davis Professorship (Israel), the first American Institute of Physics' Scientist Diplomat Award, a Titular Director of the International Union of Pure and Applied Chemistry, the Distinguished Service Award (Indiana University), an Honorary Doctorate (Eckerd College), the Distinguished Achievement Award (University of California, Irvine), and was selected by students as the Outstanding Teacher at the University of Arizona. He received his B.S. (high honors, Phi Beta Kappa) from Eckerd College and his Ph.D. in physical chemistry from Indiana University.

Ms. Loretta Peto, Secretary/Treasurer
Ms. Loretta Peto is the Founder and Managing Member at Peto & Company CPA's PLLC. She has experience in consulting on business valuation and litigation, estate and gift tax, marital dissolution and employee compensation, consulting with closely held businesses regarding business restructure, cash management, succession planning, performance enhancement and business growth, and managing tax-related projects, including specialty areas in corporate, partnership, estate and gift tax, business reorganizations, and multistate tax reporting. She is a Certified Public Accountant and accredited in Business Valuations. She is a member of the Finance Committee and Chair of the Audit Committee at Tucson Regional

Economic Opportunities. She also is a member of the DM50 and Tucson Pima Arts Council. She received a Master of Accounting - Emphasis in Taxation degree from the University of Arizona in 1984, and was awarded the Outstanding Graduate Student Award.

Dr. Janet Bingham, Member
Dr. Janet Bingham is President and CEO of the George Mason University (GMU) Foundation and GMU's Vice President for Advancement. GMU is the largest university in Virginia. Previously, she was President and CEO of the Huntsman Cancer Foundation (HCF) in Salt Lake City, Utah. The foundation is a charitable organization that provides financial support to the Huntsman Cancer Institute, the only cancer specialty research center and hospital in the Intermountain West. Dr. Bingham also managed Huntsman Cancer Biotechnology Inc. In addition, she served as Executive Vice President and Chief Operating Officer with the Huntsman Foundation, the private charitable foundation established by Jon M. Huntsman Sr. to support education, cancer interests, programs for abused women and children, and programs for the homeless. Before joining the Huntsman philanthropic organizations, Dr. Bingham was the Vice President for External Relations and Advancement at the University of Arizona. Prior to her seven years in that capacity, she served as Assistant Vice President for Health Sciences at the University of Arizona Health Sciences Center. Dr. Bingham was recognized as one of the Ten Most Powerful Women in Arizona.

Dr. Henry Koffler, Member
Dr. Henry Koffler is President Emeritus of the University of Arizona. He served as President of the university from 1982-1991. From 1982 he also held professorships in the Departments of Biochemistry, Molecular and Cellular Biology, and Microbiology and Immunology, positions from which he retired in 1997 as Professor Emeritus of Biochemistry. His personal research during these years concentrated on the physiology and molecular biology of microorganisms. He was Vice President for Academic Affairs, University of Minnesota, and Chancellor, University of Massachusetts/Amherst, before coming to the UA. He taught at Purdue University, where he was a Hovde Distinguished Professor, and the School of Medicine at Western Reserve University (now Case Western Reserve University). Dr. Koffler served as a founding Governor and founding Vice-Chairman of the American Academy of Microbiology, and as a member of the governing boards of Fermi National Accelerator Laboratory, the Argonne National Laboratory, and the Superconducting Super Collider Laboratory. He was also a board member of the Association of American Colleges and Universities, a member and Chairman

of the Council of Presidents and a member of the executive committee of the National Association of Land Grant Colleges and Universities. He was also Founder, President and board member of the Arizona Senior Academy, the driving force in the development of the Academy Village, an innovative living and learning community. Among the honors that Dr. Koffler has received are a Guggenheim Fellowship and the Eli Lilly Award in Bacteriology and Immunology.

Mr. Jim Kolbe, Member
Mr. Kolbe is a Senior Transatlantic Fellow of The German Marshall Fund of the United States. He served as a congressman in the United States House of Representatives for Arizona's 5th and 8th congressional districts from 1985 to 2007. Before joining the U.S. Congress, he served in the Arizona State Senate. He is a member of the ISGP Board of Directors and is a Senior Advisor at McLarty Associates, a strategic consulting firm. While in Congress, he served for 20 years on the Appropriations Committee of the House of Representatives, was chairman of the Treasury, Post Office and Related Agencies subcommittee for four years, and for his final six years in Congress, he chaired the Foreign Operations, Export Financing, and Related Agencies subcommittee. He graduated from Northwestern University with a B.A. in Political Science and then from Stanford University with an M.B.A. and a concentration in economics.

Dr. Charles Parmenter, Member
Dr. Charles Parmenter is a Distinguished Professor Emeritus of Chemistry at Indiana University. He also served as Professor and Assistant and Associate Professor at Indiana University in a career there that spanned nearly half a century (1964-2010). He earned his bachelor's degree from the University of Pennsylvania and served as a Lieutenant in the U.S. Air Force from 1955-57. He worked at DuPont after serving in the military, received his Ph.D. from the University of Rochester, and was a Postdoctoral Fellow at Harvard University. He has been elected a Member of the National Academy of Sciences and the American Academy of Arts and Sciences, and a Fellow of the American Physical Society and the American Association for the Advancement of Science. He was a Guggenheim Fellow, a Fulbright Senior Scholar, and received the Senior Alexander von Humboldt Award in 1984. He has received the Earle K. Plyler Prize, was a Spiers Medalist and Lecturer at the Faraday Society, and served as Chair of the Division of Physical Chemistry of the American Chemical Society, Co-Chair of the First Gordon Conference on Molecular Energy Transfer, Co-organizer of the Telluride Workshop on Large Amplitude Motion and Molecular Dynamics, and Councilor of Division of Chemical Physics, American Physical Society.

Mr. Thomas Pickering, Member

Mr. Thomas Pickering is Vice Chairman of Hills & Co, international consultants, and Strategic Adviser to NGP Energy Capital Management. He co-chaired a State-Department-sponsored panel investigating the September 2012 attack on the U.S. diplomatic mission in Benghazi. He served as U.S. ambassador to the United Nations in New York, the Russian Federation, India, Israel, El Salvador, Nigeria, and the Hashemite Kingdom of Jordan. Mr. Pickering also served on assignments in Zanzibar and Dar es Salaam, Tanzania. He was U.S. Under Secretary of State for Political Affairs, president of the Eurasia Foundation, Assistant Secretary of State for Oceans and International Environmental and Scientific Affairs, and Boeing Senior Vice President for International Relations. He also co-chaired an international task force on Afghanistan, organized by the Century Foundation. He received the Distinguished Presidential Award in 1983 and again in 1986 and was awarded the Department of State's highest award, the Distinguished Service Award in 1996. He holds the personal rank of Career Ambassador, the highest in the U.S. Foreign Service. He graduated from Bowdoin College and received a master's degree from the Fletcher School of Law and Diplomacy at Tufts University.

Dr. Eugene Sander, Member

Dr. Eugene G. Sander served as the 20th president of the University of Arizona, stepping down in 2012. He formerly was vice provost and dean of the university's College of Agriculture and Life Sciences, overseeing 11 academic departments and two schools, with research stations and offices throughout Arizona. He also served as Executive Vice President and Provost, Vice President for University Outreach and Director of the Agricultural Experiment Station and Acting Director of Cooperative Extension Service. Prior to his move to Arizona, Sander served as the Deputy Chancellor for biotechnology development, Director of the Institute of Biosciences and Technology, and head of the Department of Biochemistry and Biophysics for the Texas A&M University system. He was Chairman of the Department of Biochemistry at West Virginia University Medical Center and Associate Chairman of the Department of Biochemistry and Molecular Biology at the College of Medicine, University of Florida. As an officer in the United States Air Force, he was the assistant chief of the biospecialties section at the Aerospace Medical Research Laboratory. He graduated with a bachelor's degree from the University of Minnesota, received his master's degree and Ph.D. from Cornell University and completed postdoctoral study at Brandeis University. As a biochemist, Sander worked in the field of mechanisms by which enzymes catalyze reactions.

Biographical information of staff

Dr. George Atkinson, Ph.D.
Dr. George Atkinson is the Founder and Executive Director of the Institute on Science for Global Policy (ISGP) and is an Emeritus Professor of Chemistry, Biochemistry, and Optical Science at the University of Arizona. His professional career has involved academic teaching, research, and administration, roles as a corporate founder and executive, and public service at the federal level. He is former Head of the Department of Chemistry at the University of Arizona, the founder of a laser sensor company serving the semiconductor industry, and Science and Technology Adviser (STAS) to U.S. Secretaries of State Colin Powell and Condoleezza Rice. In 2013, Dr. Atkinson became the president-elect of the Sigma Xi Society. Based on principles derived from his personal experiences, he launched the ISGP in 2008 as a new type of international forum in which credible experts provide governmental and societal leaders with the objective understanding of the science and technology that can be reasonably anticipated to help shape the increasingly global societies of the 21st century.

Jennifer Boice, M.B.A.
Jennifer Boice is the Program Coordinator of the ISGP. Ms. Boice worked for 25 years in the newspaper industry, primarily at the Tucson Citizen and briefly at USA Today. She was the Editor of the Tucson Citizen when it was closed in 2009. Additional appointments at the Tucson Citizen included Business News Editor, Editor of the Online Department, and Senior Editor. She also was a business columnist. Ms. Boice received an M.B.A. from the University of Arizona and graduated from Pomona College in California with a degree in economics.

Marie Buckingham, B.S.
Marie Buckingham is a Fellow with the ISGP. She received her B.S. in Public Affairs with a concentration in Environmental Management and Economics from Indiana University Bloomington. Previously, she worked at King & Spalding LLP as a project assistant under the Environmental Practice Group in Washington, D.C., and also as a Sustainability Consultant to Microsoft Global in Copenhagen. She is currently applying to M.P.A. in Environmental Science and Policy programs.

Sweta Chakraborty, Ph.D.
Sweta Chakraborty is a Senior Fellow with the ISGP. She recently completed post-doctoral research on pharmaceutical regulation and product liability at Oxford University's Centre for Socio-Legal Studies and remains an active member of Wolfson College. Dr. Chakraborty received her doctorate in Risk Management from King's College London and has helped to design and co-teach a summer course in London on Managing Hazards in Europe and the United States with Indiana University's School of Public and Environmental Affairs. Her undergraduate degrees are in Decision Science and International Relations from Carnegie Mellon University.

Paul Lewis, J.D.
Paul Lewis is a Fellow with the ISGP. He worked as a Congressional Aide in Washington, D.C., and as a Legal Associate specializing in Federal Immigration Law before working with Google on Maps and Local Search products. Mr. Lewis came to Google through Immersive Media, the company behind Street View camera technology. He was involved in the rollout of Google Street View, and has managed projects involving 360-degree GPS embedded data worldwide. Mr. Lewis earned his Juris Doctor at the University of Arizona and graduated Magna Cum Laude with degrees in Journalism and Political Science from Northern Arizona University.

David Miller, M.B.A.
David Miller is a Scientific/Program Consultant with the ISGP. Previously, he was Director, Medical Advocacy, Policy, and Patient Programs at GlaxoSmithKline, where he led the company's U.S. efforts relating to science policy. In this role, he advised senior management on policy issues, and was the primary liaison between the company and the national trade associations, Pharmaceutical Research and Manufacturers of America (PhRMA) and Biotechnology Industry Organization (BIO). He also held management positions in business development and quality assurance operations. Mr. Miller received his B.S. in Chemistry and his M.B.A. from the University of North Carolina at Chapel Hill.

Raymond Schmidt, Ph.D.
Ray Schmidt is a Senior Fellow with the ISGP. In addition, he is a physical chemist/chemical engineer with a strong interest in organizational effectiveness and community health care outcomes. While teaching at the university level, his research focused on using laser light scattering to study liquids, polymer flow, and biological transport phenomena. Upon moving to the upstream petroleum industry, he concentrated on research and development (R&D) and leading multidisciplinary teams from numerous companies to investigate future enhanced oil recovery ideas

and to pilot/commercialize innovative recovery methods in domestic and foreign locations. Dr. Schmidt received his Ph.D. in chemistry from Emory University.

Ramiro Soto
Ramiro Soto is a Fellow at the ISGP. He currently is an undergraduate student at the University of Arizona College of Science seeking a Bachelor of Science degree in General Applied Mathematics. Beyond his academic curriculum, Mr. Soto is an active member of the Pride of Arizona marching band since 2010 and a member of the athletic pep band. He completed an internship with the Walt Disney Company Parks and Resorts segment in 2011. After completing his undergraduate education, he plans to apply for a doctoral program furthering his studies in mathematics.

Matt Wenham, D.Phil.
Matt Wenham is Associate Director of the ISGP. He formerly was a postdoctoral research fellow at the National Institutes of Health in Bethesda, Maryland. His research involved studying the interaction of protein toxins produced by pathogenic E. coli strains with human cells. Dr. Wenham received his D. Phil. from the Sir William Dunn School of Pathology, University of Oxford, United Kingdom, where he was a Rhodes Scholar. Prior to this, he worked in research positions at universities in Adelaide and Melbourne, Australia. Dr. Wenham received his bachelor's and honors degrees in biochemistry from the University of Adelaide, South Australia, and holds a Graduate Diploma of Education from Monash University, Victoria.

Annette M. Wetzel, M.A.
Annette M. Wetzel is Director of the Visitors Center and Special Events for the University of Nebraska–Lincoln. For more than 15 years, she has directed events for the Office of the Chancellor including academic conferences, dedications, donor events, university commencements and ceremonies. Ms. Wetzel holds a Master of Arts degree and a Bachelor of Journalism degree from the University of Nebraska–Lincoln. Her career includes an additional 15 years of experience in special event work and creative consulting in the fine paper and printing industry.